Flying Lasers, Robofish & cities of Slime

and more brain-bending science moments

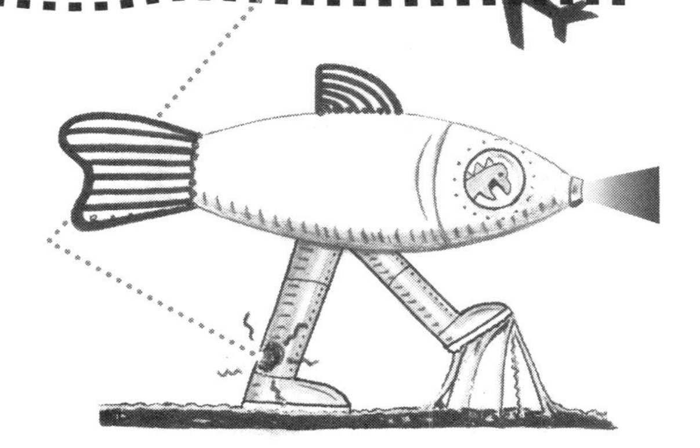

Dr

KARL KRUSZELNICKI'S

new moments in science # 2

Illustrations by Reg Lynch and Peter Pound

HarperCollinsPublishers

Books in the new moments in science series:

#1 Pigeon Poo, the Universe & Car Paint

#2 Flying Lasers, Robofish & Cities of Slime

#3 Munching Maggots, Noah's Flood & TV Heart Attacks

#4 Fidgeting Fat, Exploding Meat & Gobbling Whirly Birds

HarperCollins*Publishers*

First published in Australia in 1997
Reprinted in 1999, 2000, 2001 (twice), 2004
by HarperCollins*Publishers* Pty Limited
ABN 36 009 913 517
A member of the HarperCollins*Publishers* (Australia) Pty Limited Group
www.harpercollins.com.au

HarperCollins*Publishers*
25 Ryde Road, Pymble, Sydney NSW 2073, Australia
31 View Road, Glenfield, Auckland 10, New Zealand
77–85 Fulham Palace Road, London W6 8JB, United Kingdom
Hazelton Lanes, 55 Avenue Road, Suite 2900, Toronto, Ontario M5R 3L2
and 1995 Markham Road, Scarborough, Ontario M1B 5M8 Canada
10 East 53rd Street, New York NY 10022, USA

National Library of Australia Cataloguing-in-Publication data:

Kruszelnicki, Karl, 1948– .
Flying lasers, robofish and cities of slime.
ISBN 0 7322 5874 X.
1. Science–Popular works. I. Title. (Series: New moments in science; 2).
500

Cover illustration by Reg Lynch
Cartoons by Reg Lynch
Technical illustrations by Peter Pound
Produced in Hong Kong by Sino Publishing House Ltd on 100gsm Thailand Woodfree

10 9 8 7 6 04 05 06 07

THANKS

I would like to thank my magnificent family for being (temporarily) a single-parent family while I wrote four (4!!) books this year. (I hereby promise, in writing, that I will never be this stupid again.)

I would also like to thank Caroline Pegram (who has developed an amazing range of skills that has helped reduce my workload while improving the quality of our science communication output), Dan Driscoll (who is an inspiring radio editor and producer, and who helped shape my words into 'street' English, man), my shy-and-anonymous editor at HC (who is patient, kind, nurturing and really really clever), Reg Lynch and Peter Pound (who penned the great illustrations), Mel Calabretta and Peter Guo (who designed them into a darn good-looking book), and my two agents, Fitzroy Boulting and Paul Ricketts.

These stories would not exist without the support of the ABC, via its Program Exchange Group (PEG). For the last 10 years, PEG has commissioned me to write 670-word stories (called Great Moments in Science), which then get played every week on every ABC radio station in Australia. These short stories are the basis for the stories that ultimately appear in my popular science books.

I would also like to thank Ian Allen, Ian Vaile and Clare Byrnes at The Lab (ABC Science On-Line) for their wonderful work in building our homepage at http://www.abc.net.au/science/k2/default.htm

Finally, I would like to thank Minister Peter McGauran from the Department of Industry, Science & Technology, and the University of Sydney, for providing the basic infrastructure that allows the Office of the Julius Sumner Miller Fellow to exist.

KARL KRUSZELNICKI

BEWARE
OF THE

QUICKSAND & QUICKCLAY

If you spend a lot of time watching bad black-and-white movies on TV, you will eventually see somebody being sucked into quicksand, leaving only a hat floating on the surface.

Quicksand once swallowed up the then-slave capital of the world, a Land Rover on a beach in Australia, and an entire train on the Kansas Pacific Railroad.

There is no such thing as a type of sand which is specifically quicksand. Virtually any sand can become 'quick' if the grains of sand are kept apart from each other — usually by upwelling water.

And then there's quickclay — and it has a frightening power.

Earthquake hits wicked town

On Tuesday, 7 June 1692, on the island of Jamaica, a strange and mighty earthquake destroyed two-thirds of the harbour town of Port Royal. For a long time, historians thought that much of the town simply slid into the sea. But when a modern-day geologist actually checked eye-witness accounts of this earthquake, he came up with a new theory. George R. Clark II, a Professor of Geology at Kansas State University, claims that the town was swallowed up by quicksand!

Clark based his new theory on eye-witness reports he came across while writing a book on natural disasters. These reports on the Port Royal earthquake were written in 1694, only two years after the event. So they were probably still fresh in people's memories.

Back in those days, Port Royal was a pretty mean town and also one of the major slave markets in the world. In fact, from 1660 to 1680, the town — supposedly the 'richest and wickedest city in the New World' — was run by buccaneers. In 1674, King Charles II of Spain, with his characteristic pragmatism, sought control of the town by offering Henry Morgan, one of the most infamous pirates of all, a pardon, a knighthood, and the position of Lieutenant Governor of Jamaica. All Sir Henry had to do was just attend to the little matter of getting rid of most of his previous mates, and hanging the ones who wouldn't stay away. Apparently he didn't object.

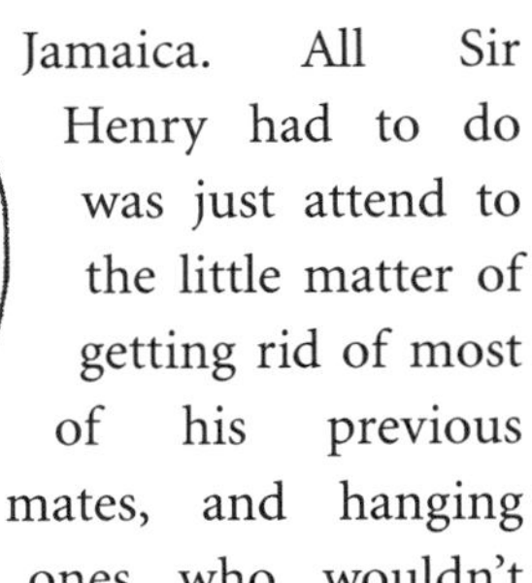

If the earthquake was God's revenge on Morgan, then He was four years too late, because Morgan died in 1688.

Port Royal was probably the worst place on the planet in which to experience an earthquake. The wicked town was built on the very tip of a 16-kilometre-long sandy spit reaching out into the harbour. Port Royal was definitely not built on a solid foundation. The sand in the top 20 metres was loosely packed and saturated with water. There wasn't anything much solid below the 20-metre mark, either — just gravels, sands, and the sporadic buried reef. Even a small earthquake would have left its mark.

We know the earthquake of 1692 was quite powerful, because, according to one eye-witness report, the two kilometre-high Blue Mountains that overlooked Port Royal were 'strangely torn and rent; in so much that they seem to be of quite different Shapes now from what they were'.

George Clark reckons that the energy from this 1692 earthquake vibrated the sands that held up the buildings in the town of Port Royal. He claims that in less than a minute, the sand had liquefied, turned into quicksand, and begun to swallow the town.

Squeezed to death

Imagine looking down on a frightened man running down one of the sandy streets of

PORT ROYAL

At the time of the great earthquake, in 1692, Port Royal was the most important city in the New World and had a bigger population (6,500 to 7,000) than Boston (6,000). It had a huge amount of shipping. In 1688, 213 ships called into Port Royal, while 226 ships called into all the ports in New England. Port Royal had safe anchorage, and deep water close to the shore.

By 1692, the town had some 2,000 buildings packed onto some 20 hectares. Many of these were made from stone and brick, and some were four storeys high.

Port Royal on 7 June 1692. After a minute or so of being vibrated by the earthquake, the sandy street begins to shine and reflect light, as the surface becomes wet. At first, his running feet on the hard sand make a flapping sound, then they make a squishing noise as the sand liquefies, and finally, a splashing sound. The previously solid sand becomes as soft as water, and the running man vanishes into the newly created quicksand.

According to the reports of the time, many people simply vanished into the quicksand never to be seen again, while a lucky few were actually carried back to the surface by an extra-strong upwelling of water currents: 'some were swallowed quite down, and cast up again by great Quantities of Water; others went down, and were never more seen'.

But a truly bizarre fate happened to those who managed to keep their heads above water. When the earthquake stopped after some six minutes (an amazingly long time — some of the most powerful modern earthquakes have lasted for only two to three minutes), the sand solidified, locking these people into the ground. They could not breathe in, due to the tight fit of the sand around them, and so they began to suffocate almost immediately.

One eye-witness said that they 'were swallowed-up by the opening Earth which then shutting upon them, squeezed the People to Death. And in that Manner several are left buried with their Heads above Ground; only some Heads the dogs have eaten; others are covered with Dust and Earth, by the people who yet remain in the Place, to avoid the Stench.' This account sounds very dramatic, but maybe even back then, reporters liked to beat up a good story!

SAN FRANCISCO, 1906

According to some reports, some buildings in the San Francisco earthquake of 1906 were literally swallowed up by the soil. If these reports are true, when the shock waves of the earthquake hit the soil used in the landfill, the soil would have behaved like quicksand. The soil then lost all of its shear strength, letting the buildings sink through it.

We know that the buildings of Port Royal didn't slip sideways into the sea, but simply sank straight down into the quicksand. For many years afterwards, the locals could row out into the bay on a calm day and see the walls of the brick buildings still standing on the ocean floor. As recently as 1859, a diver walked on some of the underwater streets and described 'the remains of 10 or more houses, the walls of which were 3 to 10 feet above sand'. That these buildings just sank into the quicksand was most probably due to the exotic phenomenon of liquefaction. (These buildings are no longer visible, due to the powerful earthquake of 1907.)

Quicksand is weird stuff

About 100 years ago, an entire train on the Kansas Pacific Railroad dived into quick-sand. Even though the workers probed down some 15 metres, they never did find the locomotive. In those days it was believed that quicksand was a special type of sand that would suck you (or a train) into the depths. But this is wrong.

Any sand can become quicksand, and anyway, it doesn't suck. Quicksand isn't matter, it's a *condition* of matter. Quicksand is just ordinary sand, but with the grains of sand kept apart, usually by upwelling water, so it behaves like a liquid. In this form, the only weight it can support is that governed by Archimedes' Principle, which states that the apparent loss in weight of a body immersed in a fluid is equal to the weight of the displaced fluid.

Imagine you're at the water's edge at the beach, standing on the hard-packed wet sand. The grains of sand are not perfectly smooth, so they lock into each other and support your weight.

But suppose that suddenly, a few metres below the surface, an underwater spring begins to gush. If the *upward* push from the current of water is greater than the *downward* suck of gravity, then the grains of sand will float freely and separately in the water. The upwelling water fills the space between the grains of sand and keeps them apart, so they can't lock together. This means that they can't support any weight. The sand then becomes 'quick', and you begin to sink.

There's another way to keep the grains of sand apart — just shake ordinary wet sand. This is what happened to the sandspit upon which Port Royal was built.

Quicksand is just ordinary sand that has water welling up through it. (In fact, if the upward flow of water is great enough, you could have 'quick gravel', or even 'quick brick'!) Not all quicksand appears wet — the Sun can dry a thin hard crust on quicksand, making it impossible to see, and very treacherous.

WHAT IS SAND?

When air, water or ice breaks rocks down into tiny particles, you end up with sand. These particles range in size from 20 to 2,000 microns (0.02 to 2 millimetres) in diameter. Everybody likes a different definition of sand, as far as the size of the particles is concerned. The US Department of Agriculture likes 0.05 to 2 millimetres, the US Army Corps of Engineers likes 0.08 to 4 millimetres, while the American Society of Testing and Materials likes 0.08 to 2 millimetres.

Quartz (silica, or silicon dioxide, SiO_2) is the most common constituent of sand. This is because silica is common in rocks and, because it's relatively hard, is almost insoluble in water and does not decompose. However, there are various sands around the world in which the major constituents are volcanic glass, iron ores or feldspar. Most sands contain small amounts of other minerals such as garnet, pyroxenes, topaz, rutile, tourmaline and zircon. Green sands get their colour from glauconite, a mineral rich in potash.

When sand is freshly 'made', its particles are usually jagged, but further erosion and weathering makes them more rounded.

In general, quicksand is caused by water being carried through rock, from a greater elevation, to emerge under a body of sand at a lesser elevation. This body of sand can be above water, or more treacherously, underwater. You generally won't find quicksand in flat country (too small a pressure gradient to separate the grains of sand), nor near gorges, ravines or canyons (the water tends to drain directly into the nearest river, or the pressure is so great that the sand is washed away).

You're more likely to find quicksand in moderately hilly country. Water can easily cut channels in dolomite and limestone rocks, so allowing upwelling springs in low places to be fed from higher places. However, dolomite and limestone do not erode to make sand, so you need different rocks upstream to provide an external source of sand.

Float, don't panic

I should point out at this stage that I have never actually seen quicksand. It's hard to get reliable accounts from people who have experienced quicksand first-hand, but I have found reports by Jim Ring (a personal experience, described in *New Scientist*, 1995), Gerard H. Matthes (engineer with the US Geological Survey around 1900, who wrote an extremely readable article in *Scientific American*, 1953), and Kirsty Irvine (a report with religious overtones, on the website of LDS Church News).

Gerard Matthes seems to have had lots of practical experience with quicksand. He says that if you fall into quicksand, don't panic — just lie on your back with your arms spread.

Quicksand is actually more dense than water (by about 1.6 times), so you will float even better than you do in water.

The first major problem to be addressed is that the viscosity of quicksand increases with shearing — so, to keep the viscosity as low as possible, you should move slowly.

Another major problem is that quicksand, unlike water, does not easily 'let go' — if you try to pull a limb out of quicksand, you have to work against the vacuum left behind.

If you have companions, they can use something (shrubs, a branch, etc.) to make a platform so they can help you out. A sheet of canvas laid onto quicksand will support a person's weight and can be used as a platform. (If a sheet of canvas is laid onto quicksand, it can even be jumped on — the quicksand under the canvas will suddenly become as stiff as rock!) Ropes would be very useful for pulling you out.

QUICKSAND BUOYANCY

You are more buoyant in quicksand than you are in water. Humans have a density of just under 1.00. Freshwater has a density of 1.00, and humans can just float in it. Saltwater is slightly more dense at 1.02, and floating is significantly easier in saltwater than in freshwater, even with such a small change in density. Quicksands have a density around 1.6, so humans float very easily in quicksand.

But if you are by yourself, and happen to be in quicksand territory, make sure that you carry a stout pole. You can get out of the quicksand by yourself, but it involves hard work. The initial goal is first to get your trunk floating, and then to get your legs floating.

First, while sinking, lay the pole upon the quicksand, and flop onto your back on top of the pole. After a while, you will have achieved an equilibrium in the quicksand, and will no longer sink. Then, with great exertion, work the pole to a new position — under your hips and at right angles to your spine. The pole keeps your hips from sinking, as you (slowly) pull out first one leg, and then the other. (Matthes recommended frequent rests during this hard work.) Once your legs are out, choose the shortest route to firmer ground, and carefully roll across the surface of the quicksand.

Rather than go off enthusiastically looking for quicksand in which to practise this technique (which you should not), you can make your own quicksand safely at home. Place a hose in a garbage bin so that the nozzle is at the bottom of the bin. Fill the bin almost to the brim with sand. Place something heavy (a dead iron, a brick, etc.) on top of the sand. Gradually turn on the water flow through the hose, and *voilà* — the object sinks. Turn off the flow of water, and the sand will again support a heavy mass. How long before Hollywood uses this concept of artificial quicksand in a movie?

Animals and quicksand

According to Gerard Matthes, different animals have different behaviours in quicksand. Most animals cannot get out by themselves.

A dog will usually get out by itself, but has to be encouraged to keep scrambling rather than to simply stay in one spot. A horse will usually get out, but has to be guided or else it will soon become exhausted.

A mule will sensibly simply lie on its belly with its skinny feet and legs underneath it, once it feels the sand beneath give way. It will usually stay quite calm and not sink at all.

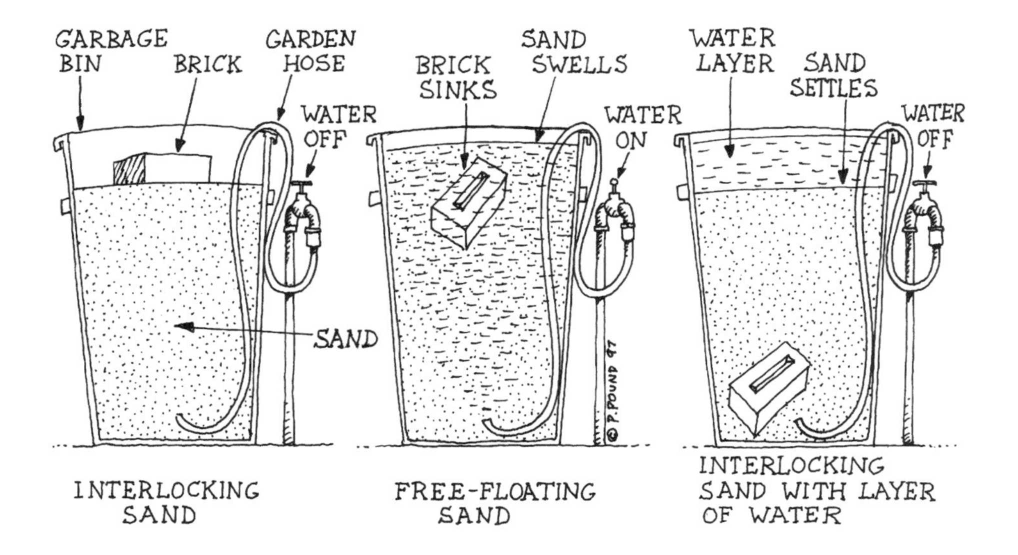

SAND & THE BIBLE

Sand is mentioned about 30 times in the Bible. In the overwhelming number of cases, it is used to compare one's friends, enemies or descendants to a very large number. For example, Genesis 32:12 says: 'But you have said, "I will surely make you prosper and will make your descendants like the sand of the sea, which cannot be counted".'

In I Kings, 4:29, sand is compared to a human quality: 'God gave Solomon wisdom and very great insight, and a breadth of understanding as measureless as the sand on the seashore.'

However, only Matthew 7:26 refers to sand for its transient qualities: 'But everyone who hears these words of mine and does not put them into practice is like a foolish man who built his house on sand.'

Cattle will almost always panic in quicksand. The only reliable way to get them out is by gently pulling them with a rope around their neck. However, once they are on dry land, they will ungratefully attack their saviours.

Unfortunately, if animals sink too deeply, their eyes will begin to bulge from their head. This is because the sand–water mix is denser than water, and so will exert hydrostatic pressure on their body. In that case they will usually have suffered so much internal damage that they will die anyway — so you may as well put them out of their misery.

Sand eats Land Rover

According to the 1993 *Guinness Book of Records*, Fraser Island, off the Queensland coast in Australia, is the largest sand island in the world. One of its sand dunes is 120 kilometres long. It has some 40 freshwater lakes, many of them high above sea-level. (As an aside, most of the trees that were used to line the Suez Canal came from Fraser Island.)

This brings me to the story I heard about how Fraser Island ate a Land Rover. A fisherman on Fraser Island, who loved fishing from the beach, had a Land Rover. It was old and tired, but the diesel engine was very rugged and economical, and the aluminium body didn't rust in its seaside environment. Unfortunately, the battery was flat. So one day, after he had started the engine with jumper cables from another vehicle, he drove along the beach to his favourite fishing spot and parked the Land Rover on solid sand, about 50 metres from the water's edge.

He didn't want to have to jump start his vehicle when he wanted to leave, so he decided to leave it idling for the hour or so that he would fish (diesels are very economical at idle).

QUICK = ALIVE

Clay that can slide is called 'quick' clay — 'quick' as in the sense of meaning 'alive'. Quickclays are called 'thixotropic', which literally means 'turning by touch' — something that, when moved suddenly, will turn from a solid into a liquid.

The fisherman had spent about three-quarters of an hour happily fishing on the water's edge, with his back to his Land Rover, when he heard the once-steady beat of the engine change to a cough and splutter. He turned around to see the tops of the wheels level with the sand — his Land Rover was half buried on what had been solid sand.

He did not know that a stream ran under the sand, into the ocean. The sand was apparently solid, but the steady vibration of the diesel engine had gradually liquefied the sand, so the Land Rover slowly sank! No shrimps on the barbie that night!

Quickclay

An example of the frightening power of quickclay occurred on 23 December 1953. A farmer called Borgen was taking a walk in the late afternoon through his farm, not far from Oslo, the capital of Norway. He saw that a tiny landslip had happened on the bank of a shallow ravine that ran through his farm. Farmer Borgen immediately rushed back home and moved all of his family into a neighbour's house. Farmers in that area knew that clay could sometimes turn into a strange stuff called 'quickclay'.

During the night, the clay on the bank of the ravine shifted, and a landslide, about 25 metres across, started moving downhill. The land upon which his house had stood suddenly became wet and slushy and began to flow through this narrow channel. In fact, it was so slushy that the avalanche kept flowing until it piled up against a highway bridge, about 2 kilometres downhill from the original tiny landslip that farmer Borgen had first noticed.

When Borgen and his family went back to their farm the next morning, all their buildings had vanished. Instead, there was now a circular crater about 10 metres deep, and over 250 metres across.

NORTHERN QUICKCLAYS

The Scandinavian and Canadian quickclays were made when the great ice sheets advanced during the most recent Ice Age — some 100,000 years ago. These clays are made of many fine materials (such as pulverised rock), which the glaciers ground up as they advanced. These fine materials were dumped on the sea floor of that time. But when the ice retreated some 10,000 years ago, these clays were uplifted and came to the surface.

One of the most damaging quickclay landslides on record occurred in 1893 at Verdal in Norway. Some 120 people died, and an area of some 5.6 square kilometres was demolished.

On 29 September 1950, in the town of Surte in Southern Sweden, there was a huge clay slide. The town had been built on deposits of clay over 35 metres deep. A new building was being built, and this occasion was the first time in Surte that a pile driver had ever been used to prepare the foundations of a building. The suspicion is that the pile driver set off the clay slide. At 8.10 am, some 3 million cubic metres of gravel and soil began to slip downhill towards the Göta River. This enormous slab, carrying 31 houses, took a railroad and a paved highway with it. The clay slide lasted for less than three minutes, but by the time it finally reached the river, it had destroyed 300 homes, killed one person and injured some 50 others.

Quebec had its turn at 11.40 am, on 12 November 1955. In less than seven minutes, the town of Nicolet 'flowed' into the Nicolet River. This clay slide carried half a dozen buildings and a bulldozer, leaving behind a hole some 10 metres deep, 120 metres wide and 180 metres long. Astonishingly, only three people lost their lives.

But clay slides also happen on the ocean floor. On 18 November 1929, an earthquake occurred south of Newfoundland. Nearby were many underwater transatlantic telegraph cables joining the USA and Europe. One after another, over a time of 13 hours and 17 minutes, a dozen of these underwater cables went out of action. A giant underwater clay slide, approximately 370 kilometres wide, travelled some 650 kilometres at a speed of around 22 kilometres per hour, snapping cables as it reached them. The telegraph engineers worked out the speed of the underwater clay slide, from the distance between the cables and the time that they one-after-the-other stopped carrying signals.

What you need to make quickclay

For clay to be 'quickclay', it needs to have four characteristics: a high water content; a low level of salt; many fine layers; and to be made up of particles or flakes less than 2 microns across.

WHAT CAN WE DO ABOUT QUICKCLAY?

Quickclays occur in many different parts of the world — Scandinavia, Canada, the Peruvian Andes and Southern Chile — and, at the moment, there is not a lot that we can do about quickclay slides. Probably the easiest and cheapest solution is take aerial photographs to identify past quickclay slides and simply not build in those areas. If we have accidentally already built towns on quickclay, we can either remove water from the quickclay, or dump salts into it. So, to prevent the destruction of some towns, we should simply dry out their feet of clay.

A high water content is the most important characteristic of wet clay. After all, when you bake clay and remove practically all of the water from it, you get bricks — which can hold very high loads. But wet clay is unstable. Sometimes observers have found small ponds left behind in the wake after a slide of quickclay. One suspected cause of the Nicolet slide in Quebec was a sewer that had broken a few months previously. Many workers thought that this sewer water could well have soaked into the clay. Similarly, the quickclay slide at Surte was also later found to be heavily loaded with water.

A low salt content in the water is another requirement for quickclay. Seawater usually has about 35 grams of salt per litre, whereas the water in quickclay usually has less than 5 grams per litre. When you dissolve salt in water, you get ions of sodium and chlorine in the water. These ions in salt help stick the particles of clay into a larger mass. When the salt leaves the clay, it turns more thixotropic.

Fine layers within the clay also help it become quickclay. Once the clay has begun to slide, they make it easy for it to continue doing so — much like the slipping of the 52 thin cards in a deck of cards.

The final requirement for quickclay is that the particles be less than 2 microns across. In most cases, these particles are crystals of various silicates, such as kaolinite, illite, chlorite and montmortillonite. Three of these clays (chlorite, illite and montmortillonite) were found in cores from the deep sea floor in the region south of Newfoundland, where the undersea slide of 1929 cut its way through a dozen underwater telegraph cables.

(A version of this story first appeared in *New Scientist*, December 1996.)

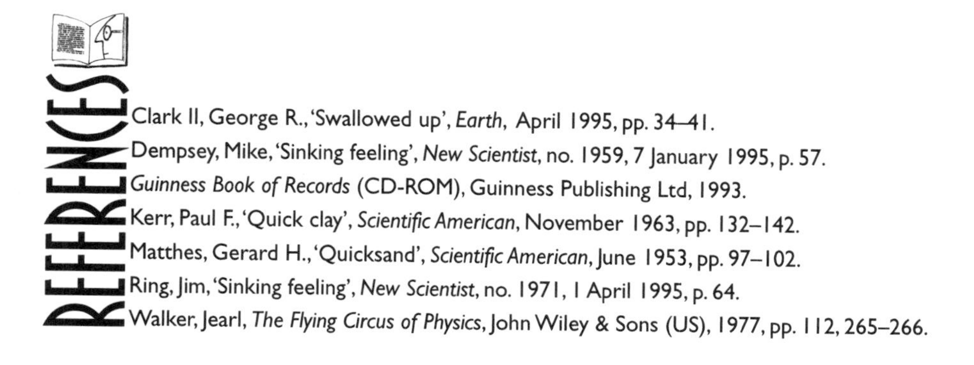

Clark II, George R., 'Swallowed up', *Earth*, April 1995, pp. 34–41.
Dempsey, Mike, 'Sinking feeling', *New Scientist*, no. 1959, 7 January 1995, p. 57.
Guinness Book of Records (CD-ROM), Guinness Publishing Ltd, 1993.
Kerr, Paul F., 'Quick clay', *Scientific American*, November 1963, pp. 132–142.
Matthes, Gerard H., 'Quicksand', *Scientific American*, June 1953, pp. 97–102.
Ring, Jim, 'Sinking feeling', *New Scientist*, no. 1971, 1 April 1995, p. 64.
Walker, Jearl, *The Flying Circus of Physics*, John Wiley & Sons (US), 1977, pp. 112, 265–266.

CANCER-PROOF, METAL-EATING, ACID-LOVING PLANTS

Plants are our friends — they are the only link that we have between the food that we put in our mouths and the energy that comes pouring out of the Sun. But plants are more than just a convenient food supply. They warn us about dangers in our environment, and they can even suck poisonous metals out of the ground.

Most of our medical drugs came from plants. And if we try to understand why plants hardly ever get cancer, maybe the lessons we learn will help human medicine.

Plants very rarely get cancer

Cancer is one of the great medical fears of the modern age. In the USA, it's the second most common cause of death after heart disease, and it kills about half a million people each year.

In Australia in 1995, cancer killed 33,805 people and accounted for 27 per cent of the total deaths in that year. One in 19 Australian men and one in 27 women will develop cancer at some stage in their lives.

But there's always been one great mystery about this thing we call cancer — why is it very rare for plants to get cancer?

Now, the first thing to realise is that cancer is not a single disease — just like infectious disease is not a single disease. There are hundreds of infectious diseases (ranging from AIDS to the common cold to measles), and there are hundreds of different cancers. For example, people talk about lung cancer as though it's a single disease, but there are actually four main different cancers that develop in the lungs. These are squamous cell carcinoma, adenocarcinoma, large cell carcinoma and small cell carcinoma. These lung cancers are different in what they look like under a microscope, how fast they grow, how you treat them, and what your chances of survival are.

The second thing to know is that any cancer is just a bunch of cells that have gone mad. So in the case of your skin, skin cells will normally grow and die, and then be replaced by new skin cells. But every now and then, one cell in your skin does not obey the rule to grow and then to die — instead, it keeps on growing. This cell can be the seed for a skin cancer.

At the moment, the usual treatment for most cancers is fairly primitive — slash and burn. The cancer specialists, or oncologists, firstly slash the cancer by cutting it with a knife to reduce it to a more manageable size. Then they burn what's left with drugs, or radiation, or both.

So when you see yet another reporter on TV talking about a new cure for cancer, what they usually mean is that for one of the hundreds of cancers, a new drug has been released that will increase the survival rate by, say, 10 per cent. What they definitely do not mean is that a 100 per cent effective cure for all cancers has finally been found.

But while cancers are fairly common in animals, it turns out that cancers generally do not happen in the plant world — even though plants are constantly exposed to high levels of ultraviolet light, which in animals causes damage to the DNA, and soon after, causes a cancer. In fact, it's very rare for a plant to get a cancer — and when it does happen, it's usually thanks to a specialised invader like a gall-forming insect or a bacteria such as *Agrobacterium.*

There seems to be a fundamental difference in the way that plants and animals handle the growth of cells, according to some recent research by Peter Doerner and his team in the Plant Biology Laboratory at the Salk Institute for Biological Studies at La Jolla, California. They fooled around with a few plants to make them have more cells.

When animals (and that includes us humans) grow any extra cells, these are normally cells that will grow without limit — in other words, cancerous cells. But when a plant grows extra cells, these cells are just spread evenly throughout the plant, and it just gets a little bigger all over.

Back when we attacked polio, we didn't build more 'iron lungs'; instead, we did basic research and eventually came up with a vaccine.

Doerner's research won't immediately give us a cure for cancer, but it will bring us closer to understanding the basic mechanisms behind this strange group of diseases.

Plants eat metals

We have all heard of the terrible nuclear meltdown that happened in Chernobyl in 1986. The countryside is still contaminated for hundreds of kilometres around. Floating in a 75-metre diameter pond, just a kilometre from the destroyed nuclear reactor, are strange rafts of sunflowers. These sunflowers are not a memorial to the tens of thousands of deaths caused by the nuclear meltdown — they're busy sucking up the radioactive metals out of the water. Yes, we're finally using plants to clean up metal pollution.

In Chernobyl, the sunflowers suck up the radioactive metals caesium and strontium. The company involved, Phytotech, from Monmouth in New Jersey, reckons that 60 sunflower plants can clean the pond in a few weeks, at a cost of about one dollar per thousand litres. The radioactive plants will be disposed of as radioactive waste. In one experiment with

POLLUTED SOIL – TRASH OR TREASURE?

Russia has never had environmental protection laws as strict as those in the Western world. The Kola Peninsula, near Norway and Finland, is hideously polluted. One of the worst areas is around the town of Monchegorsk and covers some 800 square kilometres. The Russian scientists call it a 'technogenic desert'.

In 1994 alone, Monchegorsk refineries dumped into the atmosphere some 934 tonnes of copper, 1,616 tonnes of nickel and 97,715 tonnes of sulphur dioxide.

But the factories processed ores containing tiny amounts of rare metals. Because the factories processed such vast tonnages of ore, they have dumped several hundred kilograms of platinum and palladium every year, as well as many kilograms of gold. Five kilometres south of the town, the levels of platinum and gold in the soil are 0.46 and 0.1 parts per million. These levels are roughly half the levels that can be economically extracted using conventional methods.

Each year, the factories in Monchegorsk belt out about $50 million worth of precious and non-precious elements into the environment. Most of these metals are in the top 5 centimetres of the soil. It really does seem like a job for the plants. Plants work slowly — but they work cheaply.

uranium-contaminated water in the USA, the sunflowers dropped the level of uranium from 350 parts per billion to under 5 parts per billion, in just 24 hours.

Geologists have known for a long time that certain plants prefer to grow near certain ores. For example, there are about 170 species of the plant *Alyssum.* About 50 of these species absorb nickel to concentrations of over one-tenth of one per cent of the dry weight of the plant, and some species absorb nickel to concentrations of up to 2.5 per cent. Most of the plants in this species live in Eastern Europe.

There are plants in the copper-mining belt of Zaire that absorb copper and cobalt. The alpine penny-cress (*Thalaspi caerulescens*), which lives in some parts of Britain, loves to suck up zinc.

Some plants, like the Indian mustard shrub (*Brassica juncea*), can absorb a whole pile of different metals — such as zinc, chromium, caesium, strontium and uranium. They can even absorb lead up to 60 per cent of the dry weight of the plant. The Indian mustard plant is ideal for this kind of work. It puts out shoots within two days of planting, and it grows very rapidly.

There are plans to develop large-scale filtration works based on the Indian mustard. The plants will grow in troughs hanging over the contaminated water. The roots of the plants will reach down into the water and suck up the metals. The roots can concentrate the metals to levels 100 times higher than the levels in the water. Finally, the roots will be harvested and then burnt. The carbon dioxide will go into the atmosphere, while the metals will be recycled.

Why did some plants evolve this strange ability to suck up metals, whereas most plants would be killed by such metals in the soil? Perhaps by adapting to live in metal-rich soil, these plants could live in an area that other plants could not tolerate. And perhaps the metals in the plant might act as a defence against insects that might want to have a nibble.

In the mid 1990s, the cost of cleaning up contaminated groundwater and soil in the USA was estimated to be over US$7 billion. Previous methods of decontamination left the soil dead and sterile. Perhaps the plants can help in this process.

At the moment, the plants still need some modification. Some of the plants can pick up only one metal and will die if there are other metals in the soil. Some of these metal-sucking plants grow very slowly. But perhaps genetic engineering can help us here, by transferring desired characteristics from one plant to another.

Of course, the best solution is not to put the contaminants into the soil and the groundwater in the first place.

But for those parts of the planet that are contaminated, and there are lots of them, these metal-sucking plants might be the solution. Some of us think that heavy metal sucks, but maybe that's not always a bad thing.

WHAT IS ACID RAIN?

The term 'acid rain' was first used in Manchester, England, about a century ago. Acid rain is made when oxides of sulphur and of nitrogen combine with moisture in the atmosphere to make sulphuric and nitric acids. Acid rain may fall as a liquid rain, but the term 'acid rain' is also used when the acids fall in a dry form, or as fog or snow. In some areas, such as Los Angeles or San Francisco, 'acid fogs' have been measured as being 10 times more acidic than the water version of 'acid rain'. The dry forms of 'acid rain' can also be as damaging as the wet forms.

So a more accurate name would be 'acid deposition'. Unfortunately, the term 'acid rain' has been around for a long time, and it's a bit late to change. Today, the term 'acid rain' means a precipitation (rain, snow, fog, etc.) that has a pH less than 5.7, and which was not caused by any chemical reactions involving carbon dioxide.

In many locations in Western Europe and the eastern USA, pH values of rain between 2 and 3 have been recorded.

Acid rain OK - for some plants

We've all heard about acid rain and how bad it is. In fact, in Wheeling, in West Virginia, the rain falling from heaven was once measured at a pH of 1.5, which is almost as strong as battery acid! But it turns out that some plants in Northern Europe actually need the acid rain to stay alive, and now, they're beginning to die.

The main cause of acid rain was sulphur in the coal. This coal was burnt in electrical power stations. The sulphur went up the chimney as oxides of sulphur, and came down in the rain as dilute sulphuric acid —

HOW TO MEASURE AN ACID?

You measure height in metres (or feet, if you're a pilot), and you measure weight in kilograms. And you measure how acid something is in 'pH'. (If you want to be really technical, pH is defined to be the negative logarithm of the hydrogen ion concentration, as measured in kilograms per cubic metre.)

A pH value of 7 is perfectly neutral, with absolutely no tendency to be either acid or alkaline. Anything between 1 and 6.9999+ is acid, with values closer to 1 being more acid. In the same way, anything between 7.000001 and 14 is alkaline, with values closer to 14 being more alkaline.

acid rain. Acid rain killed the trout in 20,000 of the 100,000 lakes in Scandinavia. By the early 1980s, it had also destroyed half the Black Forest in Germany.

In 1985, the problem of acid rain was so bad that many European countries agreed that by 1993 they would reduce their emissions of sulphur by at least 30 per cent. Not only did they bring in laws that power plants had to burn fuels that were low in sulphur, but they also forced them to fit equipment to their chimneys to remove a bit more sulphur.

These measures have been so successful that the amount of sulphur dumped on each hectare of land in Europe has dropped from a peak of 50 kilograms in the late 1970s, to under 10 kilograms in the mid 1990s — roughly the same level as 100 years ago.

SULPHUR

Acid rain can be caused by oxides of both sulphur and nitrogen, but the oxides of sulphur seem to be a bigger problem. Sulphur is the 16th most common element in the crust of the Earth. In the early 1990s, the world production of sulphur was about 52.7 million tonnes.

Much of this sulphur is released in industrial areas. In 1978, in the USA east of the Mississippi River alone, some 12 million tonnes of sulphur were dumped into the air from industrial sources. But most of this sulphur was 'exported' — only 20 per cent of it fell locally.

So if they have virtually stopped the problem of acid rain, how come some plants are still dying in Northern Europe?

Ewald Schnug from the Institute of Plant Nutrition and Soil Science in Brunswick, Germany, reckons it's because some plants need sulphur as an essential nutrient, and they're no longer getting it from the acid rain.

The problem of the falling crop yields is happening to a variety of plants, including grains, but it's most severe in plants in the Cruciferae family, such as oilseed rape (*Brassica napus*). This plant gives an oil that can be used in cooking and as a lubricant.

About a century ago, farmers in Europe began using new crop varieties that gave higher yields. But the plants had to be fed with more essential nutrients. So the farmers used fertilisers that contained nitrogen and phosphate, which they knew were essential plant nutrients. Although they didn't realise at the time that sulphur was also an essential plant nutrient, they accidentally used a fertiliser that contained sulphur.

And so, until the end of World War II, the plants which needed sulphur got it because the fertiliser accidentally contained sulphur compounds such as ammonium sulphate. But immediately after World War II, the ammunitions factories were closing down, so there was a vast excess in the capacity to make ammonium nitrate. This was because most explosives use nitrate chemicals. The farmers stopped using ammonium sulphate and began using ammonium nitrate in their fertilisers. Simultaneously, they also used forms of phosphate that had no sulphate in them.

This could have been a disaster for the plants, but the plants got a lucky break. There was a huge increase in sulphur in the

ACID RAIN IS BAD FOR MOST PLANTS

Acid rain has damaged nearly half of the trees in the Black Forest in Germany, as well as in many other forests. The acid 'deposition' attacks plants in two main ways. First, it leaches nutrients from the leaves. Second, it stops, or slows down, the process of the plants' roots fixing nitrogen.

In some of the mountains of Switzerland, acid rain has killed nearly half the trees. Many of these trees actually do more than just look pretty — they are a natural barrier against avalanches.

Acid rain has also killed nearly one-third of the lakes of southern Norway. If the pH of a lake falls to less than 4.8, nearly all of the molluscs and most of the fish will die.

air, because sulphur-rich coals were burnt to make electricity. So immediately after World War II, the acid rain was actually keeping certain crops in Northern Europe supplied with essential sulphur.

Some soils were already loaded with sulphur, so the plants that grew in the soils weren't really affected; and some crops didn't need a lot of sulphur, so they weren't affected either. But on average, in Northern Europe the yearly yields of oilseed rape have dropped by about 5 per cent each year since 1990. In some places where the soil was deficient and the plants

BURNING HYDROCARBONS = AIR POLLUTION

Hydrocarbons are chemicals that contain hydrogen and carbon, such as coal, alcohol and petrol. When you burn them and combine them with oxygen, the hydrogen gives you water, and the carbon gives you carbon monoxide and carbon dioxide. If only it were that simple.

***Incomplete* burning, however, can give you a bunch of other hydrocarbon chemicals. There's also the nitrogen (which makes up 80 per cent of the air) to consider. If it gets caught up in the burning process, you can get several different oxides of nitrogen.**

In the USA, hydrocarbon burning (electricity, residential heating, boilers, etc.) accounts for over 80 per cent of the sulphur dioxide and 50 per cent of the oxides of nitrogen. Some 80 per cent of the carbon monoxide comes from burning liquid hydrocarbons (such as diesel and petrol) in trucks and cars.

needed a lot of sulphur, 90 per cent of the crops have been lost.

However, there's a fairly easy but short-term cure to this problem. In northeast Germany, if farmers add sulphur to their fields of barley and wheat they can bring up the yield by as much as 20 per cent.

Who'd have thought that we'd have to compensate for a lack of acid rain!

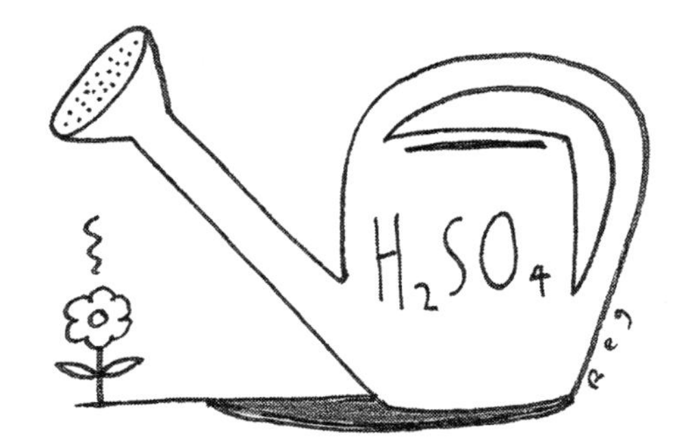

REFERENCES

Coghlan, Andy, 'How plants guzzle heavy metals', *New Scientist*, no. 2017, 17 February 1996, p. 17.

Doerner, Peter, *et al.*, 'Control of root growth and development by cycling expression', *Nature*, vol. 380, 11 April 1996, pp. 520–523.

Doonan, John & Hunt, Tim, 'Why don't plants get cancer?', *Nature*, vol. 380, 11 April 1996, pp. 481–482.

Edwards, Rob, 'Rich pickings from Russia's polluted soils', *New Scientist*, no. 2049, 28 September 1996, p. 5.

Kaiser, Jocelyn, 'Acids rain's dirty business: stealing minerals from soil', *Science*, vol. 272, 12 April 1996, p. 198.

Keeley, Jon. E. & Fotheringham, C.J., 'Trace gas emissions and smoke-induced seed germination', *Science*, vol. 276, 23 May 1997, pp. 1248–1250.

MacKenzie, Debora, 'Killing crops with cleanliness', *New Scientist*, no. 1996, 23 September 1995, p. 4.

Malakoff, David A., 'Nitrogen oxide pollution may spark seeds' growth', *Science*, vol. 276, 23 May 1997, p. 1199.

Mohnen, Volker A., 'The challenge of acid rain', *Scientific American*, August 1988, pp. 14–47.

Schnug, Ewald, cited in MacKenzie.

Simon Moffat, Anne, 'Plants proving their worth in toxic metal cleanup', *Science*, vol. 269, 21 July 1995, pp. 302–303.

CITIES OF SLIME

We humans have been looking at bacteria for over a century, but in the last 10 years we've learnt more than we did in the previous 90.

We now know that bacteria are everywhere — in the spaces between our teeth, in the sponge in our kitchen sink, and on our contact lenses. They also live 3 kilometres underground, and on the ocean floor.

It was only after we developed a special type of microscope that we could see for the first time that bacteria live in strange skyscraper cities rather than in individual mono-colonies. It was only after we had developed a new test that we could discover how many different types of bacteria there really are. The very first run of the new test revealed to us twice as many bacteria than had been discovered in the previous century! And when one scientist went looking at just one type of bacterium, he found a miniature biological version of a propeller!

A strange harmony

Our scientists have been looking at bacteria for well over a century, but only very recently have they discovered that different bacteria love to come together and live in a strange harmony, in weird cities of multi-layered slime. These slime cities are the key to really understanding bacteria.

For over a century, microbiologists have learnt about bacteria by looking at a single species growing on a glass plate that was covered with a thin layer of bacteria food. Traditionally, a few identical bacteria would be placed in the middle of the plate, then they would start to grow and divide, and grow, and so on — until they covered the entire plate.

But according to Bill Costerton from the Montana State University, less than 1 per cent of bacteria out in the real world live like this, and over 99 per cent of bacteria live in strange slime cities.

This discovery was made using a new type of microscope — the confocal scanning laser microscope. The old-fashioned microscope would let the scientists look only at dead bacteria which had been dried out into a single flat plane, and then only at close range. But this new microscope lets them look at living bacteria, and from above, from the same perspective as a plane flying over a city.

Although the new microscope was invented around 1980, it was not used to look at bacteria-in-the-wild until around 1990. And what the scientists saw looked just like a city!

Right at the bottom of the slime city was a base made from an opaque and dense slime. This base was about 5 to 10 microns thick (about one-tenth to one-twentieth of the thickness of a human hair). Towering up from this foundation were many skyscrapers, ranging between 100 and 200 microns high (about one to two times as thick as a human hair). The skyscrapers were actually bizarre colonies of different bacteria — sometimes looking like a stack of balls piled up on each other, sometimes looking like a cylinder or a cone, and sometimes looking like a mushroom. When there was a lot of food available to the bacteria, there would be lots of different skyscrapers.

REALLY, REALLY, REALLY SMALL

In our daily lives, we humans find that practically everything we deal with is roughly the same size, within a factor of a few thousand.

But the world of the scientist can be much bigger, and much smaller. In general, a scientist measures things in bands that are 1,000 times bigger or smaller than each other.

So when a scientist measures really small stuff, he or she uses 'milli' to mean a thousand times smaller, 'micro' to mean a million times smaller, 'nano' to mean a billion times smaller, and 'pico' to mean a trillion (a million million) times smaller. The next one down after that is 'femto', which means a thousand trillion times smaller.

Between the skyscrapers was a strange watery liquid, rich in many different chemicals.

This liquid was penetrated by dozens of channels or pipelines, which weaved their way between the skyscrapers. The pipelines carried everything, from oxygen and water to food and enzymes, and even garbage. But what was garbage to one species of bacteria was food to another species.

Communication

The scientists now tell us that animals communicate with each other. Back in the 1920s, Karl von Frisch claimed that bees

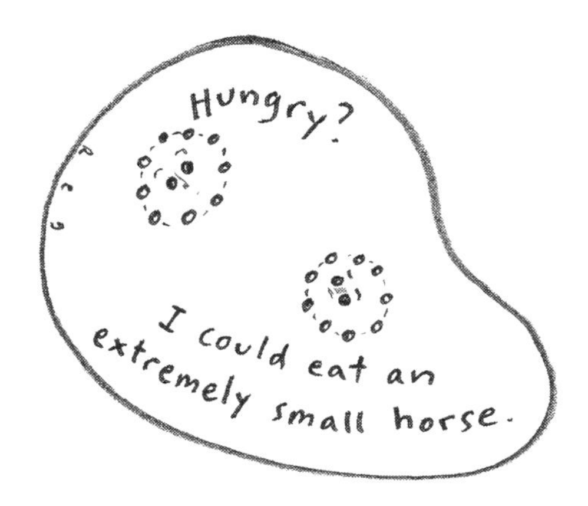

talked with each other. It took a long time to convince other scientists, so he didn't get a Nobel Prize until 1973.

Today, some microbiologists claim that bacteria talk to each other! They say that bacteria do this via various chemicals that they give off. They say that bacteria co-operate with each other so that they can, for example, fight off antibiotics, or even find and catch food.

One way that bacteria fight antibiotics is via their cell walls. Bacteria have a tough cell wall, and it's this that many antibiotics attack. But when some bacteria live in social colonies, they no longer build a cell wall, but rely on a coating of slime to protect the colony. The antibiotics can't find a cell wall to attack, and so they can't harm the bacteria.

THE DREADED GOLDEN STAPH

The bacteria called *Staphylococcus aureus*, the dreaded Golden Staph, loves to form itself into colonies. These bacteria will often form themselves into colonies on implanted biomedical devices, such as on the electrical leads of heart pacemakers. When the individual cells break off and float through the bloodstream, they are sensitive to penicillin and can be killed.

But the mother colony itself is resistant to penicillin, because it is protected by an external coating of slime. The patient with an infected pacemaker will keep on suffering recurring infections — antibiotics will not help. The only solution is to remove the pacemaker — and the colony growing on it — and implant another pacemaker (under more sterile conditions, it would be hoped).

ANTIBIOTICS KILL ONLY THE LONELY

Some bacteria have many different genes. Some of these genes are activated when they live by themselves as individual bacteria, while other genes are activated when they live in groups.

The bacteria that microbiologists have a great knowledge of are the ones that tend to live by themselves. These are the ones for which we have antibiotics. But at the moment, we have no good way of wiping out the bacteria that live in groups.

One way that bacteria co-operate to find food is to form themselves into a column, about 5 microns wide and 100 microns long. This column will ooze across the streets of the slime city until it comes near some food. It will then change direction towards the food, and then the column will break up into individual bacteria which will then eat their fill.

We now know that bacteria talk to each other. But the real breakthrough will come once we understand their chemical language. That will be the first giant step to a wonderful friendship between bacteria and humans.

Electric motors in bacteria

In 1676, Antony van Leeuwenhoek, the famous Dutch microscope maker, described one of the small creatures that he could see through one of his excellent microscopes: 'its belly is flat ... and provided with diverse incredibly thin feet, or little legs, which were moved very nimbly'. The names given to these nimble little 'legs' that Leeuwenhoek saw over three centuries ago are cilia and flagellae — and in some bacteria at least, it seems as though they actually rotate like little propellers, driven by microscopic electric motors!

Cilia (from the Latin for 'eyelashes') are the shorter of the tiny 'hairs' that grow out of the surface of a cell; flagellae (from the Latin for 'whips') are the longer ones.

Cilia and flagellae move by waving around or by spinning like a propeller, and they basically move liquid. If they're attached to a free-floating cell, like a sperm cell, they'll move that cell through liquid. On the other hand, if they're on the surface of a cell that is stuck in one place, like in your windpipe, they'll move a liquid over their surface.

You find cilia and flagellae everywhere. They're in your windpipe, moving garbage up out of your lungs. There are cilia lining the cells of the oviduct of the female mouse, to move eggs from the ovary to the uterus. There are flagellae on sperm, to help them swim — and there are modified cilia in your ears to turn sound into electricity, so you can hear.

For a long time, everybody thought that cilia and flagellae just waved around. But Howard C. Berg, who is currently at the Department of Molecular and Cellular Biology at Harvard University, didn't agree. Back in 1975, he reckoned that the flagellae

MOTORS WORK BACKWARDS

The bits of the retina (in your eye) that pick up the incoming light, and then turn it into electricity, are modified cilia.

This electricity ultimately gives you the sensation of vision.

When different cilia inside your inner ears are moved by pressure waves, they give off electricity — which ends up as the sensation of sound.

So cilia can work backwards. Not only can they turn electricity into movement, but they can turn movement into electricity.

WATER LIKE CEMENT

At our size (about 1.7 metres tall), water is a soft, and easy-to-move, liquid. But down at the size of bacteria (a few millionths of a metre across), water is actually very viscous. In proportion to a bacterium, water is as hard to move as bitumen or wet cement.

The bacteria that move around by using cilia or flagellae can't simply coast through the water — they have to keep working all the time!

that push bacteria around did not wave around in the water, but rotated rigidly like little propellers!

He's been working on this for over 20 years. In his most recent experiments, he made a hollow glass tube with a pinched-in section — so tiny that the hole was smaller than the *E. coli* bacterium he was working with.

Looking through a microscope, he sucked up a single bacterium until it was wedged in the glass pipe and unable to get past the narrow constriction. While the head of this bacterium was stuck in the glass tube, the single flagella was waving around in the open. He then bathed the flagella with a liquid containing tiny little rod-like chemicals, one of which would

SMALLEST ELECTRIC MOTOR

Japanese scientists have found a natural electric motor 10 times smaller than the natural electric motors that drive cilia or flagellae! The scientists included Masasuke Yoshida from the Tokyo Institute of Technology.

The motor is a protein called F1–ATPase, which makes a very useful energy chemical called ATP. This protein has sections that look and behave like pistons and a drive shaft.

The scientists wanted to prove that the drive shaft rotated, but how could they do this? After all, if you have ever looked at a spinning metal shaft that is smooth and has no markings on it, sometimes it's quite hard to tell if it is actually spinning. So they attached a long marker molecule at right angles to the end of what they thought was a drive shaft.

The protein now looked like it had a long skinny flag (the marker molecule) attached to a flagpole (the drive shaft). Sure enough, as the protein started to make ATP, the long skinny marker molecule began to spin around in circles. They had proved that the protein F1–ATPase was a tiny motor! This is the tiniest electric motor known to the human race.

stick to the outside of the flagella. This was a marker, so that he could see if the flagella was rotating.

And then, while the poor little bacterium was stuck inside this narrow glass tube, he fed a very low electric voltage down the saltwater inside the glass pipe and out through the other side of the bacterium — a kind of scientific cattle prod. As he fed in about one-fifth of a volt, he could see, through a video camera attached to a microscope, that the little marker was rotating about 100 times a second, or 6,000 revolutions per minute.

When he reduced the voltage, the flagella would slow down. And sometimes, when he reduced the voltage down to zero and then reversed it, the flagella would come to a complete halt and then start spinning in the other direction!

After 20 years of research, Howard Berg finally proved that some bacteria could swim around because of a tiny electric motor, spinning the flagella around like a propeller!

But even though we have proved that there is a little electric motor somewhere in the cell wall of this bacterium, no one has yet taken an electron microscope photograph of it. Since the mid 1980s or so, scientists have been getting excited about micro-machines, with tiny gears and teeth smaller than a human hair. But Mother Nature has already built machines thousands and even millions of times smaller. It's early days yet, but perhaps one day we'll have not just micro-machines, but nano-machines to work for us.

Always tiny, often useful, and many more than we ever believed!

Our planet is covered with life — animals, birds, fish, trees, plants, and hundreds of thousands of different species of insects. But when we look for life too small to see with the naked eye, we find that the textbooks say that only 5,000 different species of bacteria have been discovered. Well, the textbooks will have to be rewritten — brand-new testing techniques have discovered 10,000 previously undiscovered bacteria in a single gram of ordinary soil!

A bacterium is a living creature made up of one single cell. That single cell is a few microns across (perhaps only one-fortieth the thickness of a human hair). We humans, on the other hand, are made up of many tens of millions of different cells, each cell being roughly the same size as bacteria. We have nerve cells to think with, muscle cells to move our limbs, gland cells to make hormones, skin cells to protect us, sex cells to make more humans, and hundreds of other cell types, each with its own specialised job. But a bacterium is one single cell, which does everything.

Bacteria are incredibly useful to the human race, mainly because of the chemicals they make. There are two main types of chemicals — drugs, and everything else.

ESSENTIAL BACTERIA

Bacteria are essential for life as we know it.

Around the turn of this century, Martinus Beijerinck from The Netherlands found out that bacteria of the genus *Rhizobium* live on the roots of leguminous plants. These bacteria organise themselves into a social society that makes nitrogen and ultimately enriches the soil.

Around the same time, Sergi Winogradsky in Paris discovered that other bacteria help break up cellulose. This was an important discovery in working out the movements of the carbon in atmosphere.

One bacterium living on a golf course in Japan made a chemical that killed a parasite that infected livestock. Mevacor, a drug that lowers cholesterol, came from a sample of soil picked up by a scientist in the early 1980s in Spain.

But there are hundreds of non-drug uses of bacteria. The Danish company Novo Nordisk makes US$570 million a year by selling chemicals which it got from bacteria.

Chemicals from bacteria give us flavours and foods. They bleach paper and jeans, they process juices and even make cheese for us.

There are artificial snow fields at hundreds of ski resorts all over the planet,

BACTERIA HELP TURN GRASS INTO MILK

One classic case of bacteria-living-in-groups happens in the gut of the cow. We humans cannot digest the cellulose in grass — but cows can, thanks to five different species of bacteria.

The first bacteria are *Fibrobacter succinogenes* — they actually attack the cellulose and turn it into glucose. The glucose gets turned into another chemical, called butyrate, by colonies of bacteria called *Butyrovibrio*. A different set of bacteria then turn the butyrate into acetate. And finally, some specialised *Methanogen* bacteria turn the acetate into methane, which is then burped or farted out by the cow. But the cows need a fifth set of bacteria. These *Methanogen* bacteria are usually poisoned by oxygen, and there's a lot of oxygen in our atmosphere. So the fifth set of bacteria make a protective covering around each little colony of *Methanogen* bacteria, to keep the oxygen out.

If it weren't for these bacteria living in close co-operation, cows couldn't turn grass into milk!

and they're all based on a protein that forms ice around itself — and this protein came from a bacterium. This particular bacterium was found in a corn field by a student at the University of Wisconsin.

In the old days, bacteriologists discovered bacteria by using complicated, time-consuming and expensive methods. They would get a very small sample of something that they thought would contain bacteria, and then feed it what they thought would be the right kind of food — maybe crushed-up seaweed, or maybe ground-up ox hearts.

But the brand-new tests for finding bacteria look at the DNA and RNA of the bacteria — their genetic ladder-of-life. These tests are very fast, and very cheap.

When Norwegian scientist Vidgis Torsvik first used a DNA test on a single gram of soil, she found about 10,000 different types of bacteria — compare that to 5,000 species of bacteria discovered in over a century of bacteriology while using the old techniques!

One big issue today is 'loss of bio-diversity'. All the different (or diverse) animals, plants and insects have something to offer us. But could we ever suffer a loss of biodiversity in bacteria?

It seems almost impossible. But some of these bacteria are quite specialised, and when their home vanishes, so do they. For example, every insect has at least 100 different bacteria living on it — if that species of insect dies out, then so might the bacteria.

A green sulphur bacterium called *Chlorobium tepidum* lives in just four hot springs in New Zealand. One of these is part of a drain underneath a motel! Even though we know how to test for this bacterium and even how to culture it, it has never been found in any hot springs anywhere else in the world.

Bacteria are so useful that scientists are looking for them everywhere — in soil and in bat caves, at the bottom of the sea and in hot springs, in ancient fossils and even in Egyptian mummies.

Maybe it's a bit early to start a campaign to Save the Bacteria — perhaps we could start with Find the Bacteria!

NEW TESTS FOR BACTERIA

Thanks to advances in molecular biology, there are new tests for finding bacteria.

One very popular test looks at the ribosome, one of the many tiny biological machines inside the cell. A ribosome makes proteins. There are only very small differences between the ribosomes of bacteria of different species.

This test looks at a very small part of the 16S rRNA ribosome. In fact, according to the 16S rRNA ribosome test, humans are virtually identical to dogs and very difficult to pick from wombats. So if the 16S rRNA test picks a difference between two species, they're certainly very different. In fact, this test probably underestimates the number of species present. And best of all, because this test does not require that bacteria be fed a particular type of food they love, it's very fast and very cheap.

Underground life

Whenever we think of life, we think of something pretty close to the surface of our planet. After all, the animals run around on the surface, and most of the sea creatures spend practically all their lives within a few hundred metres of the surface. The birds do fly around in the sky, but apart from a few exceptions like the albatross (which can spend months at a time on the wing), most of the birds spend a lot of time close to the surface. But now it seems as though there is life in the very rocks several kilometres beneath our feet.

You need a few essential ingredients for life to survive. First, you need a supply of energy. Then you need various nutrients and minerals that the living creature can use as building blocks in its body. And finally, you need fluids — not just inside the body, but sometimes also to carry food in and to take waste products away.

Bacteria are really good at surviving in all sorts of unlikely places. They've been found living in vents that are 3 kilometres down on the deep ocean floor, where hot water squirts out at temperatures up to 370 Celsius degrees. They've also been found in hot springs in Yellowstone Park in the USA; in active volcanos; in giant reservoirs of oil that are 3 kilometres beneath the surface of the North Sea; and even below the permafrost of Alaska.

These recently discovered bacteria have one very important difference from practically all the other life forms on our planet — they don't depend upon the Sun for their energy.

We get our energy by eating animals and plants. The animals in turn got their energy by eating plants, and those plants got their energy from the Sun by photosynthesis. So, indirectly, we humans get our energy from the Sun. And any bacteria that live off us ultimately get their energy from the Sun.

But these underground bacteria are different. They get their energy from the chemical bonds they break when they turn big molecules into small molecules.

The other odd thing about these underground bacteria is that they seem to like it hot — much hotter than we humans can stand. We humans suffer severe protein damage once the temperature gets above 41 degrees Celsius. But some of these bacteria are quite happy living, growing and dividing at temperatures around 110 degrees Celsius! In fact, some bacteria seem to be able to survive brief exposures to temperatures around 370 degrees Celsius.

According to Karl Stetter and Angelika Hoffmann of the University of Regensburg, Germany, some of these high-temperature bacteria actually make special chemicals that stop the proteins from breaking down. These bacteria can in fact reassemble the proteins if they have been damaged by high temperatures.

Professor Lloyd Hamilton, a geologist at the Queensland University of Technology, has found microscopic remains of bacteria in solid rock. He has found them in quartz in Saudi Arabia, and in the giant McArthur River lead–zinc–silver deposit in the Northern Territory. With a microscope, you can actually see fossilised bacteria that are very similar to bacteria still alive today. Professor Hamilton suggests that in some cases, bacteria may have had a major part in laying down vast deposits of various minerals.

It seems as though our planet has a hidden biosphere — vast quantities of

KITCHEN DIRTIER THAN TOILET!

When most people think of bacteria, they think of the toilet. But when scientists from the University of Arizona in Tucson go looking, they always find kitchens dirtier than bathrooms!

The bacteria associated with human wastes are called 'faecal coliform bacteria'. The scientists hardly ever find faecal coliform bacteria when they swab the rim of the toilet, but they find them all over the kitchen.

The reason is that as soon as you have used the toilet, any droplets of water that have splashed up from the toilet bowl begin to evaporate. The porcelain surface is very smooth and there's nowhere for the bacteria to hide, and after a few hours they die.

But the kitchen sink is loaded with faecal coliform bacteria.

You might think that stainless steel is perfectly smooth, but as far as a bacterium is concerned, it's a rugged landscape with towering mountains and deep valleys, and lots of good places for bacteria to hide. Once they find a good spot, they'll start making a glue to cement themselves in position. Soon, other unrelated families of bacteria will join them. The wipe with the damp sponge won't remove the bacteria from stainless steel — it'll actually bring more bacteria in!

As long as your sponge stays wet, the bacteria can keep on living. Even if you let the sponge gradually dry out in the air, the bacteria will still survive for a few days.

Much of this work has been carried out by Carlos Emriquez and his colleagues from the University of Arizona. In one house, they found that

bacterial life hidden in the rocks beneath our feet. Bacteria certainly survive 3 kilometres underground, and can probably survive as deep as the Great Artesian Basin, which is 8 kilometres deep. The discovery of these bacteria might cause the Book of Evolution to be rewritten — maybe life didn't begin on the surface, but underground instead.

And these bacteria could give us a clue about life on Mars. When the Viking spacecraft landed on Mars in 1976, the scientists didn't come up with a clear answer about Martian life. But consider this: one of the Poles of Mars is covered by ice made of water, so maybe a kilometre underground there could be just the right temperature for life to flourish. Perhaps when the scientists examined the surface, they were looking in the wrong place.

every single surface in the kitchen that had been wiped by a sponge was heavily contaminated with faecal coliform bacteria. But on the sixth day, the surfaces no longer had germs on them. It turned out that on the fifth evening, the family had thrown away their well-worn sponge and replaced it with a new one.

Breadboards have a surface that is much more craggy than stainless steel, and so lots of bacteria live there. Cliver and Paul K. Park, from the University of California in Davis, found that you could sterilise breadboards and sponges with your microwave.

They deliberately sprinkled bacteria onto breadboards, and then cooked them in an 800-watt microwave oven. They found that a wooden breadboard took only 2–3 minutes to be bone dry and completely free of bacteria. They guessed that the bacteria probably boiled to death. However, they could not kill bacteria on plastic breadboards. They figure that this is because there's no water naturally present in the plastic to heat up and then kill the bacteria. However, you can kill all the bacteria in a plastic breadboard by running it through a dishwasher on the normal cycle.

The microwave is also handy for sterilising sponges. A wet sponge takes about a minute to sterilise, while the dry ones take only 30 seconds.

However, when I tried nuking my breadboard in my powerful microwave oven, it got so hot that it began to smoulder! So start off by nuking for shorter periods (say 3 minutes for a breadboard, and 10 seconds for a sponge), and then working your way upwards.

Already, you can buy sponges that have chemicals built in that will kill the bacteria. Similar chemicals could be built into the flat surfaces of your kitchen.

REFERENCES

Berg, Howard C., 'How bacterium swim', *Scientific American*, August 1975, pp. 36–44.

Hamilton, Lloyd & Muir, M.D., 'Precambrian microfossils from the McArthur River lead–zinc–silver deposit Northern Territory, Australia', *Mineral Deposita*, 1974, pp. 83–86.

Holmes, Bob, 'Life unlimited', *New Scientist*, no. 2016, 10 February 1996, pp. 26–29.

Raloff, Janet, 'Sponges and sinks and rags, oh my!', *Science News*, vol. 150, 14 September 1996, pp. 172–173.

Stetter, K.O., Huber, R., Blöchl, E., Kurr, M., Eden, R.D., Fielder, M., Cash, H. & Vances, I., 'Hypothermophilic archaea are thriving in deep North Sea and Alaskan oil reservoirs', *Nature*, vol. 365, 21 October 1993, pp. 743–745.

Torsvik, Vidgis, cited in Holmes.

Wickelgren, Ingrid, 'Pay dirt', *Popular Science*, March 1996, pp. 48–51.

Wu, C., 'Molecular motor spins out energy for cells', *Science News*, vol. 151, no. 12, 22 March 1997, p. 173.

Reg

ANCIENT SPEARS TO FLYING LASERS

Sometimes the power of the military can push forward a technology at an incredibly rapid speed. For example, wooden biplanes flew into World War II, and six years later we had jet planes and rockets.

Unfortunately, the ultimate aim of all military technology is to somehow harm other humans. But without spears in our distant past, we wouldn't have been so effective at killing animals for food. And in many cases, there is a flow-on from the potentially destructive military technology to a more peaceful application.

Ancient spears

One of The 10 Big Questions is: Where did we come from? We still don't have the complete answer to that enormously large question, but we are getting closer. We now have fairly solid evidence that our ancestors, 400,000 years ago, were able to plan ahead with their minds and build skilfully with their hands.

As far as we humans are concerned, evolution moved slowly for a long time and then very rapidly over a short time.

Here's the big picture. Animal life moved onto the land about 400 million years ago. The animals tended to keep one characteristic of their fishy ancestors — they had their legs out to the sides of their bodies, like the crocodiles of today. The great advantage that the dinosaurs had was the relocation of the legs to directly underneath their bodies. This kept them on top of the food chain for about 200 million years, until about 65 million years ago when (according to the theory popular in 1997) a rock about 10 kilometres in diameter smacked into the Earth and wiped them out. When this rock hit, it threw up such an enormous cloud of dust that the Earth was plunged into an almost-instant winter. The rock at the impact site was vaporised and thrown into the sky, and as it landed across half of the Earth it set fire to the trees. About 90 per cent of the trees on the planet must have burnt, according to the amount of soot found in the geological layers of some 65 million years ago. The mammals expanded into the many geological niches left vacant by the death of the dinosaurs.

About 55 million years ago, there was a creature roughly the size of a two-year-old infant and it walked on its hind legs. This creature evolved into several different types of bipeds. About 3 million years ago, one of those bipeds — our direct ancestor — had a brain half the size of the modern human brain and was the height of a young teenager.

Then an incredibly fast evolution happened. The brain size doubled in just 3 million years. Our ancestors of 2,500 years ago had the same size brains as we have today. They had access to much less 'knowledge' than we have today, but they were just as 'smart' as we are now. Indeed, 2,500 years ago, the astrologers and astronomers of Babylon were able to predict solar and lunar eclipses. Back then, they knew of the 18-year cycle (actually 6,585⅓ days) called the Saros, which is essential to the prediction of eclipses. I know that I personally can't predict eclipses, and so I have a great deal of respect for their intelligence. But at what stage between the 3-million-year-old

THE FIRST WEAPON

The spear, as a sharpened stick, was probably the earliest weapon we humans ever used against an enemy or an animal. This pole-with-a-sharp-point-at-one-end can be held in the hands while thrusting, or it can be thrown from a distance. In 3200 BC, in Mesopotamia, armies were already using spears. By 1000 BC, the Greeks had modified the spear into a pike 2–3 metres long.

The sarissa, an extra-long pike 4–6.5 metres long, was introduced by Philip II of Macedon around 350 BC. The advantage of this extra-long pike was that you could poke the enemy Greeks before they could poke you with their shorter pikes. The sarissa was one of the main weapons used by Alexander the Great when he forged his huge empire.

The Romans in turn modified the spear into the pilum, a javelin about 2 metres long.

Around the 14th, 15th and 16th centuries, the masters of the pike were the Swiss, and various European monarchs used them as mercenaries.

The spear was also modified into the lance, to be used by soldiers on horseback. The last significant use of the spear or lance was in September 1939 — at the beginning of World War II — when brave Polish cavalrymen charged German tanks, on horseback.

ancestor and ourselves did 'intelligence' (whatever 'intelligence' is) appear?

One thing that humans do is fight, and the spear (the long pointed stick) is probably one of the first weapons invented. The earliest proof that we have of our ancestors killing an animal lies in the ribs of a mastodon (an early type of elephant), which was found in Ohio. This mastodon had died 12,000 years ago, and there are clear marks of spears on the ribs.

But it now seems that the story of spears goes back even further.

In 1995, three 400,000-year-old spears were discovered in a 6-square-kilometre open-cut brown coal mine in Schöningen, Germany. The spears were found close to stone tools, and were surrounded by thousands of horse bones (many of which showed signs of being butchered). These spears are 2 metres long and, just like modern Olympic javelins, their centre of gravity is at the forward one-third of their length! These spears were cut vertically from a 30-year-old spruce tree, with the tip of the spear carved from the denser wood near the base of the tree! These two facts — the location of the spears' centre of gravity, and the fact that the spears' tips were carved from denser wood — tell us that 400,000 years ago our ancestors had intelligence, depth of planning, knowledge passed-down from generation to generation, the patience needed to make tools, and a high order of tool making.

The good news is that we are getting slowly closer to answering the 'Where did we come from?' question. The bad news is that the 50-metre-high rotary shovel excavator at Schöningen, with its 11-metre-high cutting wheel that can cut through several tonnes of coal each minute, is still churning through the brown coal — and through possible evidence of our ancestors.

So while we're digging for our future, we're destroying evidence of our past.

The sniper

To the average person, a sniper is a skilled shooter, usually attached to the military, whose job is to spot and kill enemy soldiers whilst hiding in a concealed place. The bullets seem to come from nowhere, one after the other. Snipers are usually incredibly demoralising to the enemy. There have been cases where a single skilled sniper has been able to stop the advance of an entire battalion. But now some recent technology should be able to destroy the sniper's advantage — by sniffing out the incoming bullet.

HISTORY OF SNIPING

The first sniper activity of which we have records dates back to the 12th century, when the Welsh used their longbows to defend themselves against the Norman invaders.

The Royal Marines have a rather exact definition of the job of the sniper. They say that the job is 'to stalk, locate and observe a well-concealed target and destroy personnel and equipment with precision fire, using the full capabilities of weapons, optics and ammunition'.

Around AD 1450, the first gunsights appeared on rifles to improve their accuracy. In 1737, King Frederick the Great of Prussia wrote how he had used a rifle with a telescopic sight in a target shoot. In the American War of Independence in the 1770s, American backwoodsmen were regularly sniping British officers.

The word 'sniper' actually comes from the bird, the snipe. This smallish mottled bird has a long skinny sensitive beak, which it uses to feel for grubs and worms in the swamps. The words 'sniper' and 'sniping' were invented in the late 18th century by British officers serving in India. They reckoned that snipes were a very good game bird, because when they're flushed out of the grass they go into a twisting zigzag flight which makes them harder to shoot. The British officers would use camouflage and inconspicuous movements to sneak up on the snipe and kill it — and that's how the word 'sniper' came into the English language.

There have been many infamous and famous snipers in history, such as Lee Harvey Oswald, who supposedly shot President John F. Kennedy, and James Earl Ray, who supposedly shot Martin Luther King Jnr. In the Sniping Hall of Fame, which you can find on the World Wide Web, Marine Gunnery Sergeant Carlos Hathcock is rated as one of the greatest snipers of all time. In Vietnam, he was credited with 93 confirmed kills and another 200 unconfirmed kills. And there have been many cases in which an offender who was holding a few hostages at gunpoint has been killed by police snipers so surgically that the hostages were unharmed.

At the moment the top three sniper rifles in current use are the Heckler & Koch PSG–1, the Accuracy International L96A1 and the Barrett model 82A1. The Barrett is enormously powerful and punches the bullets out of the muzzle at a speed of 853 metres per second. Even at a range of 1,500 metres away from the sniper, the bullets from the Barrett can easily punch through a thick brick wall.

Even today, with all the high-tech weapons used on the battlefield, the sniper is still one of the most feared soldiers. The sniper will usually shoot his bullet from a well-camouflaged site, knocking off key targets such as radio operators and officers, before invisibly fading back into the countryside. But now there are technologies that make it harder for the sniper.

One approach uses three passive sonar sensors that listen for the sound that the bullet makes as it punches through the air. The sensors use this information to calculate, almost instantly, the path, height, angle of travel, speed and size of the incoming bullet. The sensors will even tell you the distance that the bullet will miss you by — if you're lucky. This system is totally passive, and does not use an emitted signal like radar to find the bullet. This means it can't be detected by the enemy.

Another system is being developed by the Lawrence Livermore National Laboratory in California. Not only can it track incoming bullets, it can do the same for missiles and artillery shells. It doesn't listen for the sound that the bullet makes, but rather looks for the heat that it gives off as it pushes through the air.

With these new technologies, snipers will be stripped of their most valuable asset — invisibility. As soon as the sniper fires a bullet, this new technology will tell where the bullet came from. So the sniper's first shot might be his last.

Flying lasers

During the Gulf War back in 1991, the technologically superior Allied Forces found themselves unable to deal with the cheap and nasty Iraqi Scud missiles. Sure, the Allied Forces managed occasionally to hit the incoming Scud missiles with their own anti-missile missile, the Patriot. But quite often the many smaller pieces of debris from the combined explosion of the Patriot and the Scud caused more damage than if the Scud had simply been allowed to travel through its full path and land where it was aimed.

There had to be a better way — and it turned out to be a big fat laser!

Ever since the laser was invented back in 1960, people have put it to all kinds of uses — cutting and measuring in industry, measuring air and water pollution in scientific research, slicing away diseased parts of the body in medicine, sending information at enormous rates in communications, identifying groceries at the cash register in a supermarket, and even making 3-D holograms in the arts.

The military have always lusted after the laser as a weapon. In fact, during the Cold War back in March 1983, President Reagan proposed that a program be started to build an anti-ballistic missile system. This so-called Star Wars system would use giant lasers down on the ground, which would bounce beams off so-called 'fighting mirrors' in orbit, to blast missiles out of existence.

They were a very tempting target. The liquid-fuelled missiles had thin skins, and if you could just play a powerful-enough

GERM-FINDING LASER

The Caliope laser was invented to look for evidence of the manufacture of chemical or biological warfare agents. This recently declassified American laser can actually tell the difference between bacteria at a distance of up to 18 kilometres!

The Caliope laser squirts out a beam that bounces off the germs, and then the complex reflections in the returning light are analysed 'by computer beings (also called genetic algorithms) whose job it is to analyse data and mimic the job of evolution'. Even though it's only a prototype, this system can already pick the difference between bacteria such as *E. coli* and *Streptococcus*.

It has already successfully analysed the light from factory smokestacks to work out what products are being made inside. While the prototype Caliope laser works well, it is too large to fit into an unmanned aerial vehicle or a satellite. But within five years, the system should be small enough to fit.

Consider the peaceful medical uses! Imagine being able to play a low-power laser beam onto enlarged tonsils or onto the hands of a patient — and be able to diagnose instantly what bacteria are present.

beam on the one spot it would burn a hole through in a few seconds.

But there were too many problems.

First, most lasers are quite weak — only a few thousandths of a watt, and you need millions of watts to burn through metal.

Second, our atmosphere is not perfectly smooth, but has turbulence in it. If you have ever flown, you have felt turbulence. It's actually caused by 'cells' or pockets of air having a slightly different temperature from the surrounding air. These cells can be a metre across, and they have two bad effects on a big fat laser beam — not only do they widen and weaken the beam, they actually bend it away from the fast-moving target.

Third, how do you find the tiny, fast-moving target?

Fourth, how do you aim the beam so accurately that it will heat up just a tiny part of the incoming missile — and from a distance of several hundred kilometres?

It was all much too hard, so, in 1993, after spending US$30 billion on Star Wars, the project was wound down. But technology marches on, and now the US Air Force reckons that they can stuff all this capability into a 747 jumbo jet. They have awarded a billion-dollar contract to Boeing, Lockheed-Martin and TRW to build an Airborne Laser, and the first flight is scheduled for the year 2002.

TRW has come up with a laser that can generate 400,000 watts. A dozen of these lasers strung together can squirt 5 million watts at the target. The laser cleverly recycles environmentally friendly chemicals, so it is possible to carry enough fuel on a 747 to fire off 30 laser shots, each lasting about 5 seconds. Each shot will use about US$1,000 worth of chemicals.

Lockheed-Martin has come up with a clever way to find the missile by first looking for the hot rocket exhaust, then 'painting' the missile with weak laser beams and analysing the reflected light to find the missile, then finally letting off the big fat killer laser beam.

But what about the turbulent atmosphere? Well, Lockheed-Martin has worked out a way to find out exactly what kind of turbulence there is between the 747 and the target, and can compensate for the turbulence in the atmosphere by changing the shape of a mirror thousands of times per second and then bouncing the laser beam off that mirror.

By 2008, seven of these airborne fortresses should have been built, at a cost of around US$5 billion. Light is made up of photons, so, just like in *Star Trek*, the captain will say: 'Fire off the photon torpedos!'

Perhaps this powerful offspring of Star Wars will have a peaceful use. If it can blast out an energy beam at the speed of light, maybe it can deflect incoming asteroids!

History of radar

In 1900, Nikola Tesla, the Yugoslav-born American scientist who invented the modern AC induction motor, became the first person to talk about the possibility of using radio waves to find moving objects.

In 1904, the German scientist Christian Hulsemeyer used radio waves to pick up ships at a range of 2 kilometres. He patented his device, believing that it could be used to stop ships colliding at sea.

The next step was taken in 1922, with a collaboration between the Italian physicist Guglielmo Marconi and the American inventors A.H. Taylor and L.C. Young. They realised that if you knew the speed at which the radio waves travelled, and if you could accurately measure the time that it took the radar signal to come back to your antenna, you could work out the distance to the object. By 1932, the US Naval Research Laboratory was able to pick up planes from a distance of 80 kilometres.

In 1936, the Scottish physicist R.A. Watson-Watt used a magnetron to amplify the radar impulses being transmitted from the radar unit. He also used an amplifier to boost the signal of the echoes. These improvements were incorporated as the Radio Detector Telemeter in the chain of radar stations along the British coast. The Americans made their own improvements on this device in 1938, and called their unit the SCR–270.

By 1938, a collaboration between Telefunken, the Luftwaffe and Wurzburg had successfully tested a 560-megahertz radar unit.

In 1939, at the outbreak of World War II, France, Japan, Germany, the United Kingdom and the United States were all doing research into radar.

The Germans had already got their first radar unit, the 125-megahertz Freya, working for the Maritime Signal Office. Goering was directing research into radar at some 200 German Institutes and eight Military Research Centres. But Germany's great mistake came in 1940, when Hitler, who had absolutely no scientific education, banned all research into electronics, because he said that electronics was a 'Jewish science'.

But in the early days, the new technology of radar wasn't highly trusted. On 7 December 1941, the Opana Radar Station of the US 55th Signal Aircraft Warning Service near Kahuku Point in Hawaii picked up a strange radar signal. The privates on duty interpreted this signal to mean that they were picking up incoming aircraft, at a range of 220 kilometres. They passed this information on to their commanding officers, who told them to ignore this signal as they must be from US planes coming from the mainland. In fact, the incoming aircraft were the first of 353 Japanese carrier-based war planes that sank or damaged 18 warships, destroyed 188 aircraft on the ground, and killed 2,330 servicemen and 100 civilians.

By this time, the United Kingdom was already well ahead of everyone else in radar. British scientists had developed the cavity magnetron, which could transmit a very narrow beam of radar waves. This meant the exact location of approaching aircraft could be determined very accurately. The United Kingdom was also the only military power to have radar units small enough to be carried on planes. This meant that British fighters could find German bombers in the darkness of night.

To try to protect its military advantage with secrecy, the British Air Ministry put out a bizarre story, claiming that the reason the British night fighters were so successful in seeing and shooting down German bombers was because of a regular diet of carrots!

There is some degree of truth in this story. Carrots are rich in carotene, which our body turns into vitamin A. And vitamin A is essential for the manufacture of 'visual purple' — pigment used by the rods in the eye to see at low light levels. A balanced diet will give you enough vitamin A from foods such as green vegetables and

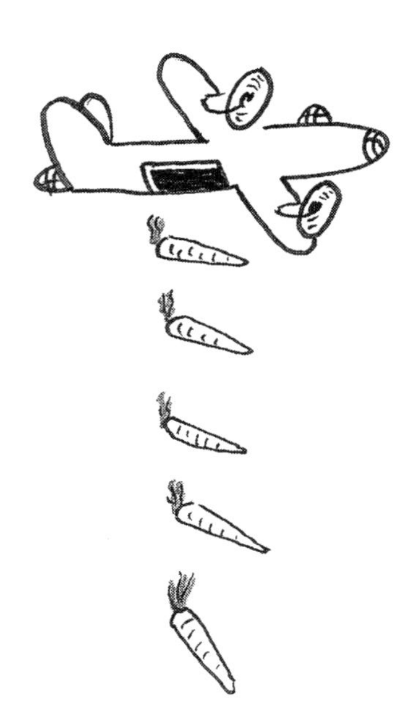

A from foods such as green vegetables and dairy products — extra vitamin A will not make your night vision any better. But for a while at least, the story fooled the Germans.

The real reason that the British pilots were so successful was that a chain of hundred metre-high radar masts had been built along the European coasts of England. This chain could detect German bombers some 160 kilometres away.

In 1943, the Germans shot down a British plane carrying top-secret radar equipment. Suddenly the Nazis realised how the British had been able to shoot down Nazi bombers at night, and reactivated their research into electronics — but they were too late.

Micro-radars

If you have ever been near your local airport, you're almost certain to have seen at least one radar dish in your life. Complete radar systems can be very large and very expensive. One of the largest and most expensive radar installations ever built was part of the United States Ballistic Missile Early Warning System. When it was installed way back in 1960, in Greenland, it cost US$500 million. But radar seems set to move into the home, with a complete radar unit the size of a matchbox and costing only US$10!

Radar stands for Radio Detection And Ranging. 'Radio Detection' means finding something, such as a ship or aeroplane, with radio waves, while 'Ranging' means finding its distance. A radar unit sends out radio waves, and then listens for the echoes. It measures the time for the echo to return and so it can work out the distance to the object.

The micro-radar story began when Tom McEwan was working as an electronics engineer at Lawrence Livermore National Laboratory in California. He was measuring incredibly short pulses of energy emitted from nuclear reactions. So he built hundreds of tiny electronic gates that could open and close in less than a billionth of a second. He used this technology to make micro-radar units.

His micro-radar unit emits a million pulses per second, with each pulse being less than a billionth of a second long. It radiates less than a millionth of the power of a cellular phone, so it's regarded as perfectly safe to humans. It uses so little power it will run on two AA batteries for

RADAR IN SPACE

In 1942, during the Nazi bomber attacks, the British radar operators noticed strange interference on their radar screens. At first they assumed that the Germans had begun to jam the British radar units. However, they were actually picking up the first radio emissions from the Sun ever detected.

In fact, some 10 years earlier, the American radio engineer Karl Jansky had also picked up radio waves from space. He had been trying to find out what had been interfering with commercial radio signals. He was a good experimentalist, and he noticed strange radio signals that reached a maximum loudness every 24 hours, as the Milky Way came above his antenna. He correctly thought that this was because radio signals were coming from the centre of the Milky Way.

Hydrogen (chemical symbol H) exists in vast cold clouds in the space between the stars. In 1945 the Dutch astronomer Van de Hulst said that radio signals, at a wavelength of 21 centimetres, should be able to be detected from this hydrogen. These atoms of hydrogen were not exactly active — in fact, the average atom of hydrogen floating in space would emit a tiny burst of radiation only once every 11,000 years!

But space is very big, and there are a lot of hydrogen atoms out there. In 1951, Van de Hulst's predicted radio signal was detected. It turned out that the hydrogen was arranged in a fairly thin layer around the centre of the galaxy.

The hydroxyl (chemical symbol OH) molecule was found in 1963, on the 18-centimetre wavelength — also in clouds around the centre of our galaxy.

It was only as recently as 1970 that scientists finally found the combination of H and OH (H_20, or water) floating in space. They also found a fairly complicated molecule called formaldehyde, and the new science of Interstellar Chemistry began. Around this time, people began to think about whether life could have formed from molecules floating in the vast clouds in the spaces between the stars. Today, Interstellar Chemistry is a rich field with hundreds of chemicals so far detected in space.

Soon another new science was invented — Radar Astronomy. This is the science of studying various bodies in the solar system by bouncing radar signals off them, and analysing the echoes. The Moon was the first to be mapped, with Mars and Venus close behind. We cannot see the surface of Venus, because it's shrouded by thick layers of cloud. But not only was radar astronomy from Earth able to crudely map the surface of Venus, but it was also able to measure its rotation. On Venus, a 'day' takes 243 Earth-days, but a year takes 225 Earth-days — so on Venus, the day is longer than the year!

But radar astronomy works only out as far as Saturn. Beyond that, the distances are so great that the reflected signals are currently too small to analyse.

eight years! Even so, his micro-radar can transmit and receive radar waves over a distance of some 60 metres.

The new micro-radar is, according to Tom McEwan, 'turning out to be a Swiss Army Knife'. Lawrence Livermore Laboratories has already received over a thousand commercial enquiries, and has issued over 30 commercial licences.

The first commercial cheap-radar product can exactly pinpoint steel reinforcing rods buried in concrete. But micro-radar could find vertical timber studs behind your plaster walls, and it might even find structural defects.

The car makers want to use this micro-radar as a collision-avoidance device, to help you while parking, or even to look in your blind spots while you are driving on the highway.

It would be ideal at measuring fluid levels. The brewers of beers can't use ultrasound to measure beer levels in their vats, because the foamy froth gives such a good echo. But radar waves can look through the froth to see the top of the liquid. And radar waves could measure the levels in your fuel tank, radiator and oil sump. Washing machines could use radar waves to sense how much clothing had been loaded, and then add the appropriate amount of water and detergent.

The medical profession can see hundreds of uses for cheap radar units. Radar waves could see through your shirt and skin to 'see' the actual movements of your heart muscles. Radar waves could also 'see' the movement of your lungs. This could lead to machines to protect against Sudden Infant Death Syndrome. The micro-radar could easily pick up the baby's breathing, even through a mattress.

At the moment, this miniature radar costs around US$10 and is about the size of a box of matches. But when large-scale production begins and the radar shrinks down to a chip about the size of your little fingernail, it could cost about 10 cents!

Just 30 years ago, the laser was a very interesting but useless laboratory device. Today you'll find lasers everywhere, from your hi-fi system to your local supermarket. The same could happen to the micro-radar. But such a device could be a two-edged sword. On the one hand, this radar-on-a-chip will have thousands of uses. But on the other hand, it will add, albeit slightly, to the sea of electronic noise and smog that already washes around us.

REFERENCES

Ballantyne, Ian, 'Spot the killer!', *Focus*, October 1996, pp. 42–45.
'Bullet detector', *Popular Mechanics*, August 1996, p. 20.
Dennell, Robin, 'The world's oldest spears', *Nature*, vol. 385, 27 February 1997, pp. ix, 767–768.
Fulghum, David A., 'Boeing team tapped to build laser aircraft', *Aviation Week & Space Technology*, 18 November 1996, pp 22–23.
'Not your average calliope', *Aviation Week and Space Technology*, 31 March 1997, p. 19.
Scott, William B., 'Sniper detector revived for airborne use', *Aviation Week and Space Technology*, 26 August 1996, p. 64.
Thieme, Hartmut, 'Lower Palaeolithic hunting spears from Germany', *Nature*, vol. 385, 27 February 1997, pp. 807–810.
Watkins, Steven, 'The destroyer rides out', *New Scientist*, no. 2055, 9 November 1996, pp. 38–40.

MAYAN MYSTERY & MISTAKEN AGRICULTURE

Agriculture is a bit like mother's milk — everybody thinks it has to be good for you. But Professor Michael Archer, from the University of New South Wales, claims that agriculture has been the major cause of the degradation of many ecosystems in Australia. He compares the relative areas devoted to agriculture and mining, and says that agriculture is 3,500 times more destructive to the environment than mining. Maybe so, but you can't eat rocks!

Every major civilisation has depended on agriculture, and the Mayans were no exception. They were a single political group as early as 2000 BC, but their Golden Age was very short — AD 250 to AD 900. And then

suddenly, their society collapsed. This collapse has been one of the great mysteries of archaeology, but one theory claims that it could have been caused by something as simple as a change in the weather.

Mayan mystery collapse

The Mayan culture began to flourish around AD 100 on a long skinny strip of land between North America and South America.

The Mayans created a magnificent civilisation. They didn't have a single central government, but rather a whole bunch of independent city-states that were governed by kings. Most of the time, these city-states lived in peace and traded for goods freely with each other. As time went by, the architecture in their cities became more elaborate, with bigger and fancier pyramids. The cities were fed by farms that grew cocoa, cotton, corn, tobacco and fruit. The cities covered square kilometres, and had population densities higher than have ever been achieved since in that region.

The Mayans were fascinated with time and astronomy. They didn't invent either telescopes or clocks, but they could predict eclipses of the Sun and the Moon. They were fascinated by the planet Venus, and their calculation of the time that Venus took to make a complete loop through the sky was accurate to within two hours of the real figure of 583.92 days.

They used the calendar to work out the best times for planting and harvesting

MAYAN CALENDAR & END OF WORLD

The Mayans believed so deeply that their calendars could predict the future, that, in AD 1695, one of their rulers decided that they should surrender to the invading Spanish! He claimed it was predicted in the calendar. So in December 1695, the Mayans let the Spanish know that they thought they would lose in battle. This battle happened on 13 March 1697. The Mayans lost, as they predicted!

There are actually two Mayan calendars: a 365-day solar year and a 260-day sacred year. These combine to give a complete cycle running for some 52 solar years, before previously used dates reappear. The Mayans firmly believed that disasters (major and minor) happened every 52 years.

According to their calendar, the end of the next cycle will occur on 23 December 2012 — interpreted by some gullible people to be the day on which the world will end. As you probably guess, I don't think it will.

crops, for trading or hunting animals — and when best to declare war or to negotiate peace. Their calendar was similar to ours, with 365 days making up each solar year. The Mayan calendar has as its starting date 2 August 3114 BC.

The Mayan calendar follows a cycle. The Mayans firmly believed that what had happened in the past would happen again in the future. And according to their calendar, the end of the next cycle will come on 23 December 2012.

Around AD 800 the Mayan's sophisticated empire began to crumble, and within 200 years it had almost vanished — and we don't know why. But according to research by Dr David Hodell and his colleagues from the University of Florida in Gainesville, the biggest drought for 8,000 years hit the Mayan civilisation around AD 800.

He and his team measured the temperature 1,200 years ago by looking at cores drilled out of a lake bed on the Yucatan peninsula. They looked very closely at the *water* in the cores.

Water is made of hydrogen and oxygen. There are two isotopes of *oxygen* — a heavy one and a light one. So you get two isotopes of *water* — a heavy one and a light one. Water evaporates when it's warm, but the

ISOTOPES

There are some 92 different elements that make up the Universe as we know it. But many of these elements exist in more than one variety. These varieties are called isotopes.

A conventional model of the atom is that it looks a bit like a solar system — with a core at the centre, and a bunch of electrons in various orbits whizzing around the core. The core is made up of positively charged protons, and neutrons (which have no charge).

Hydrogen, the simplest element of all, always has one proton. This proton gives the core a single positive charge, which is balanced by the negative charge on a single electron whizzing around the core.

There are three different isotopes of hydrogen. The core of a hydrogen atom can either have just a single proton (in which case it's called hydrogen), or a single proton and a single neutron (in which case it's called deuterium), or a single proton and two neutrons (in which case it's called tritium).

Hydrogen, deuterium and tritium are all different isotopes of the element that we call hydrogen. These three isotopes are called 1H, 2H and 3H.

A proton and a neutron weigh about the same, and each is roughly 2,000 times heavier than an electron. So tritium is roughly three times heavier than ordinary hydrogen, while deuterium is nearly twice as heavy.

Oxygen has three isotopes: oxygen 16, oxygen 17 and oxygen 18. They all have eight protons in the core of the atom, but have varying numbers of neutrons: 8, 9 or 10. As 18O is about 12 per cent heavier than 16O, 16O is more likely to evaporate in hot weather than 18O — so there will be more 18O left behind in the mud.

lighter water always evaporates more easily. And the hotter it gets, the more of the lighter water evaporates, leaving the heavier water behind. If you look at the core, and find more heavy water than normal, you know that the climate was hotter at that time.

But of course, it takes more than just a severe drought to knock over a civilisation which once reached all the way from the Yucatan peninsula in Mexico, through Guatemala and Belize, and down to Honduras. It turns out that around the same time, the cities were getting very large, putting severe environmental stress upon the farms. Also around this time, the quality of the farming land had seriously been degraded by how the Mayans did their agriculture. And to really increase the stress on their society, the various city-states had begun to have wars with each other.

The theory that the Mayan empire collapsed when triggered by a severe drought led Jeremy A. Sabloff, of the University of Pennsylvania Museum of Archaeology and Anthropology, to ask this question: 'How severe do the internal stresses in a civilisation have to become, before relatively minor climate shifts can trigger widespread collapse?'

So you have to wonder what a long hot Greenhouse summer would do to our highly stressed society.

Agriculture – worst mistake in history?

Was agriculture the worst mistake we humans ever made in our entire history? Would we have been better off had we stayed hunter–gathers?

Nowadays we in the wealthy countries have a pretty good style of living. Most of us have easy access to clean water, good education and high-quality medical care, and many of us can get some of the material goodies — like a colour TV, or a second-hand car. The current theory is that all of this wealth became possible, once the human race left behind hunting and gathering and turned instead to agriculture.

You might think that you need agriculture to give you a stable social base, so that the arts and the sciences can give us Leonardo da Vinci and Beethoven, Scud missiles and digital watches. But remember, those great carvings and cave paintings in the Australian Outback and France were created by hunter–gatherers up to 40,000 years ago.

A hunter–gatherer supposedly stores no food and grows no plants, so every day begins with a new struggle to get more food.

But the Kalahari Bushmen of Africa spend only 19 hours each week getting food. Each person has over 2,000 calories and 90 grams of protein a day, which is more than they need. And the Bushmen have access to some 75 different wild plants, giving them a very well-balanced and diverse diet.

In Europe some 10,000 years ago, the hunter–gatherer societies were not nomadic, but lived in fairly permanent villages usually near bodies of water. They had well-planned foraging techniques so that, by hunting and gathering both animals and plants from water and from land, and by taking account of what the different seasons offered, they were able to provide quite a steady and diverse food supply.

Unfortunately, when societies changed from hunting and gathering to agriculture, the quality of life actually went down! Agriculture could support a larger population, but the population was less healthy. Some 10,000 years ago, the hunter–gatherers in Greece and Turkey had an average height of 170 centimetres. But once their societies had adopted agriculture, their heights dropped 15 centimetres — or 6 inches in Imperial measurements. Even today, Greeks and Turks are still not as tall as their ancient ancestors!

We see the same pattern in American Indian skeletons in Illinois and Ohio. Around AD 1150, the hunter–gatherer society gave way to one based on farming corn and maize. The difference between the hunter–gatherers' and the farmers' skeletons tells a sad story. The farmers had four times more iron-deficiency anaemia, three times more infectious diseases, 50 per cent more enamel defects in the teeth (indicating poor nutrition), and a drop in life expectancy from 26 years down to 19 years.

There are three main reasons why agriculture is bad for your health — at least

ENERGY FLOW IN AGRICULTURE

In 1963, Roy A. Rappaport tried to work out the energy costs of agriculture in New Guinea. He looked very closely as the Tsembaga turned some 100 square metres of secondary forest into a garden. They had to clear the underbrush, clear the trees, fence the garden, weed and burn in the garden, place soil retainers on the edges of the garden, and then plant and weed until the end of the harvest.

He then measured the weight of crops that they raised, and compared this to the amount of energy that they had to put in.

It turned out that the Tsembaga got quite a good-term yield on their energy investment. When they grew taro-yam their yield-to-input ratio was 16.5:1, and for sweet potato it was 15.9:1.

So they certainly got back much more energy than they put in. From the point of view of energy, agriculture is marvellous.

MEAT IS MARVELLOUS?

The !Kung-San of the Kalahari Desert in Africa value meat much more highly than plants as food. In New Guinea, people spend an awful lot of time raising pigs, even though they have lots of carbohydrate plants available. And many hunter–gather societies have a special word to indicate that they are 'meat hungry'.

Meat is a rich source of proteins, but it needs some fat with it. If you don't eat lean meat with other calories, the amino acids in the meat will be converted into energy, rather than into proteins to build up your body. It is a habit among the Pitjandjara of Australia to inspect the tail of a killed kangaroo for fat, and then to leave the animal to rot if it doesn't have enough fat.

in its early days. Firstly, those ancient farmers had a fairly restricted diet, which was not as healthy as the diverse diet of the hunter–gatherers. So while they usually had enough calories, they were a bit short on proteins, minerals and vitamins. Secondly, as the farmers became more dependent on only a few crops, a failure of just one of those crops could virtually wipe out their society. And finally, because so many people lived together in one small group in an agricultural society, infectious diseases spread rapidly.

But agriculture also led to terrible social changes — like a loss of equality and a loss of freedom.

Everybody has to work in a hunter–gatherer society, the food is shared between all members of that society, and there's no great difference between the leaders and the followers.

But in an agricultural society, you can have one group of farmers who make the food, and another group who steal food from the farmers and grow fat on the proceeds. This is a very common pattern in history. Greek skeletons from Mycaenae show that members of the royal family some 3,500 years ago were much better off than the commoners — they were 6 centimetres taller, and had six times fewer cavities in their teeth. The upper-class Chileans of AD 1000 had four times less bone damage than the working class.

So why did the hunter–gatherers take up agriculture? The answer seems to be a combination of climate change and increasing population.

About 10,000 years ago, the planet gradually came out of a 100,000-year-long Ice Age, and the climate was a bit variable. So the pattern of the seasons changed

unpredictably, and this really upset a society that lived off wild plants and undomesticated animals. A society would have had a better chance as it tamed animals and rapidly selected and bred plants.

Also, the population got bigger. So the human race sacrificed certain advantages — a small population, exercise, a healthy diverse diet, and equality — for a large population that ate more empty calories, and that suffered more disease during a shorter life in a non-democratic society.

Now, as the 20th century draws to a close, about half of the available workforce on the planet is involved in agriculture to try to feed our enormous population. Even so, some two-thirds of our people don't get a well-balanced diet. We are using about one-quarter of our available arable land to do this agriculture. Unfortunately, there isn't enough free land to go back to hunting and gathering. So we can't go back to hunting and gathering, and anyway, hunting and gathering could not resist a changing climate and an increasing population of farmers, and it had to fade out. Life is not a Rambo movie, and one healthy hunter–gatherer could not hold off 100 unhealthy farmers.

REFERENCES

Diamond, Jared, 'The worst mistake in the history of the human race', *Discover*, May 1987, pp. 64–66.

Harris, Marvin, 'The 100,000-year hunt', *The Sciences*, January–February 1986, pp. 22–32.

Headland, Thomas N., 'Paradise revised', *The Sciences*, September–October 1990, pp. 45–50.

Hodell, David A., *et al.*, 'Possible role of climate in the collapse of Classic Maya civilisation', *Nature*, vol. 375, 1 June 1995, pp. 391–394.

Rappaport, Roy A., 'The flow of energy in an agricultural society', *Scientific American*, September 1971, pp. 117–132.

Sabloff, Jeremy A., 'Drought and decline', *Nature*, vol. 375, 1 June 1995, p. 357.

Reg

El Niño & the Line in the Pacific

Every now and then, in different parts of the ocean, currents suddenly well up from the ocean floor, bringing nutrients to the surface. These upwelling currents affect less than 0.1 per cent of the surface of the oceans, but they give us over 50 per cent of all the fish caught. These short-term changes give us the El Niño and a strange line in the Pacific.

The line in the Pacific

The story goes that the only artificial object visible from space is the Great Wall of China. Like many myths put out by the advertising agencies, it ain't true. You can't even see the line of the Great Wall of China from a jet plane at 15 kilometres up. But in the Pacific Ocean the astronauts have seen, with the naked eye, a living line that moves to the West at 2 kilometres per hour!

The astronauts have seen a few artificial phenomena, such as the lights of the tuna boats near Japan, the enormous patterns caused by irrigation schemes in the northwest of the USA, and the marked contrast in vegetation on each side of the bunny fence in the southwest of Western Australia. But the Great Wall of China?

Well, it's a bit like looking from the top of a building and trying to see a fishing line on the ground. It doesn't matter how long the fishing line is — if it's too skinny, you just can't see it.

You see, an average human eye should be able to pick up an object smaller than one minute of arc — or, in plain English, you should be able to see something about a millimetre thick at a distance of 3 metres. At the height of the space shuttle, about 300 kilometres, you should be able to see something about 100 metres wide — twice the length of an Olympic swimming pool. The Great Wall of China is narrower than this, and anyway it is in low contrast, being grey stone on a grey background.

But this living line in the Pacific is about 2 kilometres wide. It was photographed because of the Joint Global Ocean Flux Study, a 10-year-long international program to study the ocean. The scientists wanted to look at how various currents and chemicals in the ocean moved around the planet. So they used ships, aeroplanes, satellites and Space Shuttles, and also drifting platforms deep under the surface. They took a wide range of biological, chemical and physical measurements, looking at factors such as ocean colour, movement of carbon, movements of warm and cold currents, surface temperature and so on. They even used Orion P–3 aircraft flying at 150 metres to radiate the water with a weak blue-green laser beam, which made the chlorophyll in the water fluoresce. Chlorophyll is important because it's the main chemical in phytoplankton that turns sunlight into energy.

One of the photographs taken from the Space Shuttle clearly shows an almost-straight line, hundreds of kilometres long, in the Pacific Ocean. Going back through ships' records over a century or so, it turns

ANCIENT ENORMOUS MASSES OF DIATOMS

The Line in the Pacific is currently just a temporary phenomenon. It happens every few years, and lasts for only a short time. But according to drillings of the Pacific Ocean floor, these growth spurts of diatoms may have lasted for thousands of years in the past!

Scientists have found there have been at least five periods in the last 15 million years where it seems as though enormous mats of diatoms, covering an area the size of Australia, floated in the Pacific Ocean and then sank to the floor to build up layers some half a metre thick. They happened about 15 million years ago, 12.5 million years ago, 10 million years ago, 6 million years ago and the last one about 4.5 million years ago.

We're not sure what caused them, but we do know that North America joined up to South America at Panama around 4 million years ago. Maybe some change in the global circulation of water around our planet meant these vast mats of diatoms no longer sprang into existence.

UPWELLING CURRENTS

Moving air can push coastal water out to sea. This warm water is then replaced on the coast by colder water welling up from the ocean floor. This cold water is very rich in nutrients such as nitrates, phosphates and silicates. Plankton eat these nutrients, breed and reproduce very rapidly, and in turn, get eaten by bigger creatures, such as fish.

In fact, while these areas of 'upwelling' make up less than 0.1 per cent of the surface area of the oceans, they provide more than half the fish caught in the world.

out that this temporary line in the Pacific tends to happen between August and the following January. Around this time, cold water, loaded with minerals and other nutrients, upwells from the ocean floor until it hits the surface, and then begins rolling to the West at about 50 kilometres per day. At the surface, there's a zone about 2 kilometres wide where a slab of cold water 70 metres thick dives under a slab of warmer water 40 metres thick.

The water can be quite turbulent in this narrow zone, with breaking whitecaps. But there are also patches of pale green water just loaded with a tiny algae, a diatom called *Rhizosolenia*. These diatoms are about two to three times thicker than a human hair. In 1926, the naturalist and explorer William Charles Beebe said that when the line passed his ship, these diatoms were 'so abundant that in places they were of the consistency of soup'. It's the colour that makes the mixing zone visible from space, and the colour comes from the combination of the whitecaps, the colder water, and the squillions of diatoms.

Diatoms are a type of plankton. The word 'plankton' comes from the Greek word *planktos*, meaning wanderer. Plankton just float and wander around the oceans of the world, carried by the currents. Plankton are very important because these tiny floating plants make 90 per cent of our oxygen. Diatoms have silicon bodies, and when they die and sink they turn into diatomaceous earth. Human uses for this diatomaceous earth include fertilisers, swimming-pool filters, soundproofing materials and even paint removers.

When the cold water wells up from the ocean floor, these diatoms pile up just ahead of the cold front as it moves to the West and they feed on the nutrients in the cold water. According to the measurements taken in the Joint Global Ocean Flux Study by the laser in the low-flying aircraft, these diatoms were 100 times more abundant in the zone between the cold and hot water than elsewhere. But not only were there lots of diatoms having good feed, they were also having sex — once a day, they were dividing and making baby diatoms. It seems as though once a pool of cold water wells up from the ocean floor, this leads to a very localised and very productive food web which lasts for only a very short time — but the diatoms make hay while the Sun shines. And the bigger creatures which eat diatoms also know when to show up.

So it turns out that a collection of fish, foam and future fertiliser, when all lined up, fighting, feeding and fornicating, are more visible from space than the Great Wall of China!

El Niño

There's another phenomenon, also related to upwelling currents, which has Australia's farmers at its mercy. It helped the invading Spanish Conquistador, Francisco Pizarro, wipe out the Incas in the 1500s. It's that disastrous weather phenomenon called El Niño, and even at the end of the 20th century there's nothing we can do about it.

This whole El Niño thing is a bit complicated, and mathematicians and climatologists are still trying to work it out. But you can get a pretty good idea of it if you just remember a few things.

The first thing to remember is that, just like a pot of water on a stove, you get more water vapour rising from warm water than from cold water. If you're talking about the Pacific Ocean, this means you get more rain clouds above warm water than above cold water. The second thing to remember is that wind can push water around. The third thing to remember is that wind flows from high pressure to low pressure. The final thing to remember is that on our planet, practically everything is linked to practically everything else — so South America and Australia are linked by a giant pond called the Pacific Ocean.

So here we go. The normal state of affairs in the southern parts of the Pacific Ocean is that the pressure is high over Tahiti and low over Darwin — so the winds flow from Tahiti to Darwin. This link between the air pressure around Tahiti and Darwin was discovered in the early 1920s by Indian mathematicians working for the British. There's a kind of seesaw effect — in other words, an oscillation — when the pressure goes up in one place, it goes down in the other. The meteorologists call this the

WHY DARWIN AND TAHITI?

Sir Gilbert Walker, Director-General of Indian Observatories in 1904–24 and later Professor of Meteorology at the Imperial College of Science and Technology, tried to predict the monsoons. The monsoons are very important for India because they influence very strongly India's agriculture. At that time, there simply was no good Theory Of Weather — so he tried the brute-force approach of fishing through every piece of data he could find!

In the early 1900s, Indian mathematicians were very cheap, very plentiful and often very clever. He gave these mathematicians every piece of data on weather that he could get from the entire British Empire, as well as from everywhere else in the world. The breakthrough came in 1924, when one of his mathematicians realised that every time the barometric pressure went up in Darwin it went down in Tahiti, and vice versa.

Suddenly Sir Gilbert realised that there was some sort of linkage of atmospheric pressure systems across the Pacific. He called this oscillating pressure system the Southern Oscillation.

Southern Oscillation, because it happens in the Southern Hemisphere.

So thanks to the high pressure in Tahiti, the trade winds normally push warm Pacific Ocean surface water up against the east coast of Australia. In fact, the water level can be 40 centimetres higher on the Australian side of the Pacific than on the South American side. (Underneath the warm water, there's a layer of cold water reaching right down to the ocean floor. Off Australia, this cold water is about 200 metres under the surface, while off South America it's only about 50 metres under the surface.)

Rain clouds form above the warm water, and so it's usually fairly wet on the Australian side of the Pacific Ocean.

The dry air that has been stripped of water rises and then flows back across the Pacific to South America, and this dry air sinks onto the coastal region of South America. This means no rain.

But there's another reason why there's very little rain on the South American coast of the Pacific. In Peru, for example, the coastal waters are cold, and there's not much evaporation and not much rain. So, under normal conditions, the west coast of Peru is a harsh, treeless strip of desert some 100 kilometres wide and 2,000 kilometres long. (There's another reason as well — the peaks of the Andes stop the moist air from the forests of the Amazon getting to the coast.)

But it's not all bad for Peru — the cold ocean waters are loaded with food and well-up just off the Peruvian coast, so the fishing is great. In fact, about a quarter of a century ago that tiny area off the coast of Peru

HOT POOL IN PACIFIC

Part of the normal state of affairs in the Pacific Ocean near the Equator, is that there is an enormous pool of warmish water on the west side of the Pacific. But in an El Niño-Southern Oscillation (ENSO), as the strong easterly winds weaken, this pool can shift 5,000 kilometres to the East.

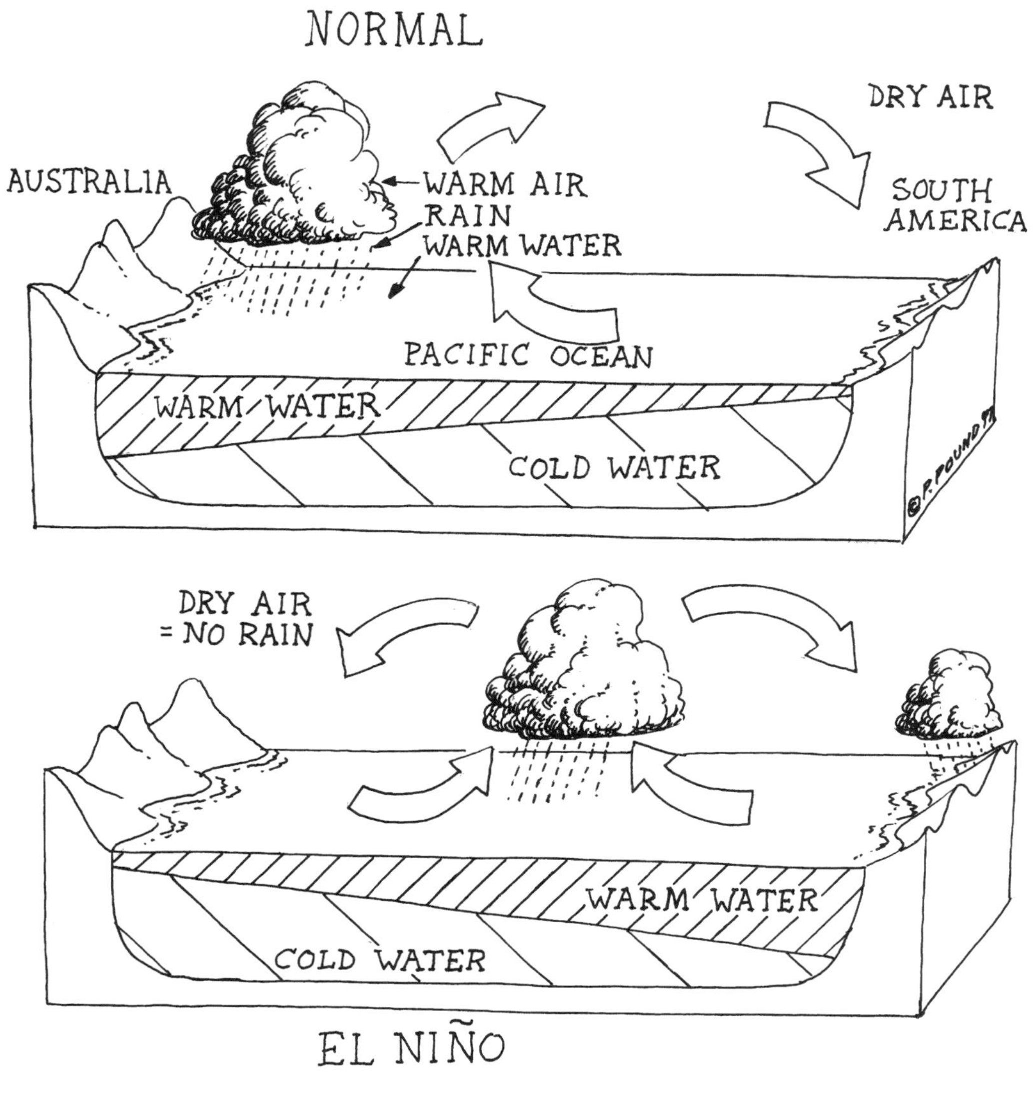

accounted for 22 per cent of all the fish caught in the whole world!

Normally, around October or November each year, the air pressure *drops* in Tahiti and goes *up* in Darwin — and so the winds start to flow the other way, to South America. The winds push the warm water over to Peru. It takes about two months for the warm water to arrive — usually around Christmas time. On the one hand, this warm water (up to 7 Celsius degrees hotter than normal) stops the nutrient-rich cold water from welling up, and so there is a temporary and minor fall in the fishing catch. On the other hand, the warm water means that there is a bit of rain on the Peruvian coast. And in Australia, the cold water off the coast means less evaporation, which means a temporary lull in the rains.

In most years, everything comes back to normal after a few months. But every now and then, this temporary state of affairs (with the warm water off Peru) lasts for many months, or even years — and we call this event an El Niño. We had an El Niño

CONFUSING NAME 1

The name 'El Niño' is actually American Spanish for 'The Christ Child' — *El* means 'the', while *Niño* comes from the Old Spanish *Ninno* and means 'child'.

In the 19th century, the Peruvian fishermen would use the name 'El Niño' to refer to the annual appearance, around Christmas time, of warm water off the coast of northern Peru and Ecuador. This warm water was just a part of the regular southward-running current that would appear around Christmas, and fade away by March. This warm current would normally flow just a few degrees south of the Equator. This warm water caused a temporary and minor fall in the fishing catch but after a few months everything would come back to normal.

Today we use the name 'El Niño' to refer to the event when this warm water reaches further south, hangs around for a year or so, and stops the upwelling of the cold water that is loaded with nutrients.

CONFUSING NAME 2 – AIR & WATER

El Niño refers to a current in the water that comes around every year or so off the coast of Peru. 'SO' is short for 'Southern Oscillation', which is a change in the normal air pressure difference between Darwin and Tahiti.

When both of these events happen around the same time, you have that terrible weather disaster that happens every two to seven years or so. 'El Niño-Southern Oscillation' is a bit of a mouthful to say, so it's usually shortened to ENSO. This is the proper name to use when you want to discuss that distressing event. Unfortunately, most people in the news media call it the 'El Niño', which confusingly is the same name used for the regular annual event.

CONFUSING NAME 3

El Niño has two meanings. To the Peruvian fishers, it's something that happens every year. But to Australians, it's something that happens every two to seven years and leads to a drought.

year in 1982–83, another one in 1986–87, another one in 1991–92, and a weak one in 1993.

The El Niño event of 1982–83 was a worldwide disaster — Melbourne had a dust storm, Australia had the worst drought in a century and the value of its rural produce dropped by A$2 billion, Africa starved, and in India the life-giving monsoons failed to arrive. On the other side of the Pacific, areas that hadn't had a cyclone for 60 years had six cyclones in six months, the rainfall in Peru increased 400-fold, and water and flood damage on the coasts of North and South America cost about US$2 billion, central Chile had disastrous rainfall and floods, and cyclones reached as far East as Tahiti.

In fact, centuries before, it was an El Niño that helped wipe out the Incas. The harsh arid desert coast of Peru is usually very inhospitable, and very difficult to cross. But an El Niño event, with its rain, can make the normally dry coast quite lush and easy to cross.

The first El Niño on record happened in 1525–26, when Pizarro was able to cross the hundred-kilometre-wide coastal desert of Northern Peru — only because, in the words of R.C. Murphy, 'he chanced upon the desert shores during one of the rare *años de abundancia*, or years of abundant water and vegetation'. Once Pizarro made it across the normally inhospitable coastal desert and explored inland, he could see that the Incas had an enormous and very wealthy empire. He returned with 180 men and 37 horses. He was able to wipe out the Incas — thanks to his deceitfulness, his superior weapons technology, and the civil war that divided the Inca Empire.

So now you understand that the pool of warm water, which is normally off the coast of Australia, sometimes runs downhill to South America because the trade winds stop blowing onto Australia's coast. The rains come from the rain clouds that form above the warm water — so when the warm water goes to Peru, so do Australia's rains. And when this happens for a *year* or more, instead of just a *month* or two — that's what we call an El Niño event.

There's not much Australia can do about the weather, but it can do something about

KELVIN & ROSSBY WAVES

As part of the El Niño, the warm water on the east side of the Pacific Ocean heads for the coast of the Americas. This very slow-moving 'wave' is called a 'Kelvin wave'. Once it hits the coast of the Americas, it is 'reflected' back across the Pacific Ocean. It is now called a 'Rossby wave', which is also very slow moving.

The Rossby wave can take 10 years to cross the Pacific Ocean at the level of Los Angeles, but 30 years at the latitude of Portland, Oregon! These waves have a 'memory' of their past El Niño, and in their own right can cause further weather changes. For example, when oceanographers studied the Kuroshio current off the coast of Japan, using the data from satellites, they found that the Rossby wave of the 1982–83 El Niño had actually pushed the Kuroshio current further North, warming up the northwest Pacific!

Right now, the Rossby waves of the El Niños of 1986–87 and 1991–93 are moving across the Pacific Ocean.

its economy. The mix of Australia's exports is very similar to the mix of a Third World country's exports — lots of primary produce, and not much in the way of value-added goods. As long as this continues, Australia will always be at the mercy of the elements.

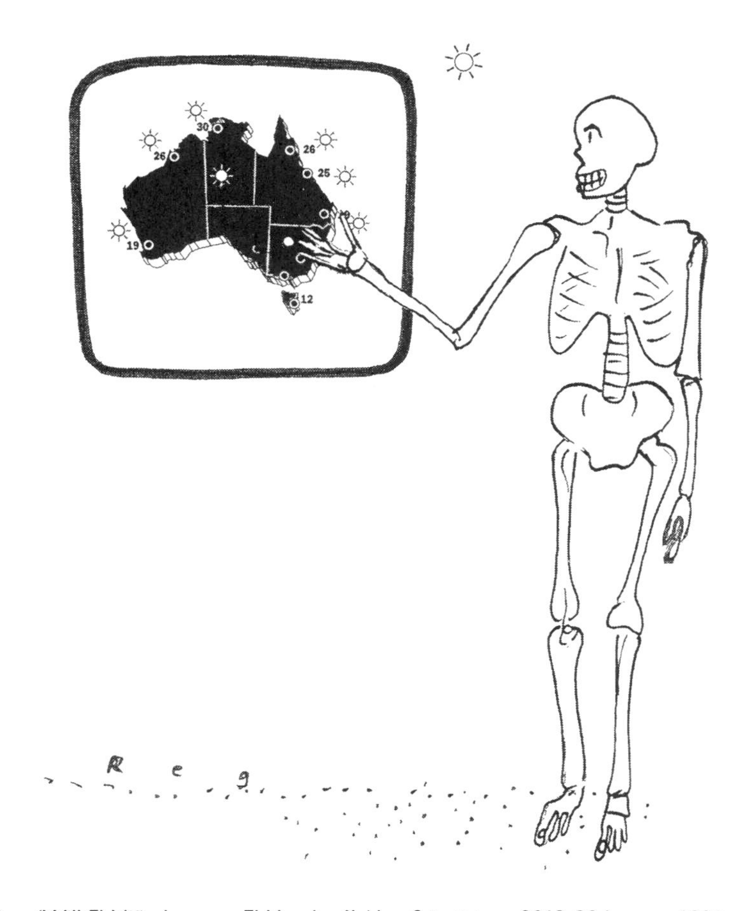

REFERENCES

Bergeron, Lou, 'Will El Niño become El Hombre?', *New Scientist*, no. 2013, 20 January 1996, p. 15.

Chelton, Dudley B. & Schlax, Michael G., 'Global observations of oceanic Rossby waves', *Science*, vol. 272, 12 April 1996, pp. 234–238.

Kemp, Alan E.S. & Baldauf, Jack G., 'Vast neogene laminated diatom mat deposits from the eastern Equatorial Pacific Ocean', *Nature*, vol. 362, 11 March 1993, pp. 141–144.

Murphy, R.C., 'Oceanic and climatic phenomena along the west coast of South America during 1925', *Geographic Review*, vol. 15, 1926, pp. 26–54.

Picaut, J., *et al.*, 'Mechanism of the zonal displacements of the Pacific warm pool: implications for ENSO', *Science*, vol. 274, 29 November 1996, pp. 1486–1489.

'Pool moves', *Science*, vol. 274, 29 November 1996, p. 1441.

Ramage, Colin S., 'El Niño', *Scientific American*, June 1986, pp. 55–61.

Stone, Roger C., Hammer, Graeme L. & Marcussen, Torben, 'Prediction of global rainfall probabilities using phases of the Southern Oscillation Index', *Nature*, vol. 384, 21 November 1996, pp. 252–255.

Yoder, James A., Ackleson, Steven G., *et al.*, 'A line in the sea', *Nature*, vol. 371, 20 October 1994, pp. 689–692.

CHARGE!
R c g

ICEBERG ARMADAS, TIBETAN ICE-BLOCK & RUSTY OCEAN

Our planet has been orbiting around our Sun for about 4.5 billion years. For most of that time, our planet has been free of ice. In fact, for most of our history there hasn't even been any ice at the North or South Poles. But every now and then, the ice comes raging down from the Poles towards the Equator. In fact, we've discovered that our climate can flip into, or out of, an Ice Age with frightening speed — and very soon (any time in the next few thousand years), we might flip into the next Ice Age!

We've discovered that we can actually change the climate by dumping carbon dioxide into the atmosphere and setting off a temporary Greenhouse Effect. Can we reverse this Greenhouse Effect by removing carbon dioxide from the atmosphere? One oceanographer has claimed he could suck carbon dioxide out of the atmosphere and into the oceans, by sprinkling the oceans with iron! Another oceanographer has claimed that Tibet has acted as another device to suck carbon dioxide out of the atmosphere.

But while we humans pop in and out of life on a 70-year cycle, the Ice Ages come and go on their million-year cycles.

Iceberg armadas

The term 'Ice Age' is a bit confusing. In fact, there are two kinds of Ice Age: the 'Big' ones, and the 'Little' ones. 'Big' and 'Little' refer to the *timing* of the Ice Ages, not how much of the planet they cover.

The Big Ice Ages come around roughly every 150 million years. The last Big Ice Age happened about 150 million years ago. Before that, there were Big Ice Ages about 300 million years ago, about 430 million years ago, and before that, about 680 million years ago. These Big Ice Ages lasted between 1 million and 20 million years.

A Big Ice Age is just a series of Little Ice Ages. The Little Ice Ages each last for about 100,000 years. There can be anything between 10 and 200 Little Ice Ages in the thing that we call a Big Ice Age.

During our current Big Ice Age, which began about 2 million years ago, the ice caps grew and advanced from the North and South Poles towards the Equator, and then retreated. Later, they advanced again and retreated again, and so the cycle repeated. On average, each advance lasted about 100,000 years, and each retreat lasted around 20,000 years.

The most recent Little Ice Age wound down about 15,000 years ago. We might roll into another Little Ice Age in a few thousand years, or we might have come to the end of this cycle — at the moment, we just don't know.

It's possible that Big Ice Ages might be related to events within our galaxy.

It also seems very likely that the Little Ice Ages are controlled by something in the geometry of our Solar System. Back in 1924, the Yugoslavian scientist Milutin Milankovitch put forward his theory. His theory is pretty successful at explaining why, during one of these Big Ice Ages that happen every 150 million years, our planet will oscillate in and out of a series of Little Ice Ages.

In our Solar System, all the planets, including the Earth, go around the Sun in a roughly circular orbit, and they're all roughly in the same plane. Milankovitch

150-MILLION-YEAR GAPS

We're not really too sure why there is a 150-million-year time interval between the Big Ice Ages. However, some scientists think it might be something to do with the fact that our galaxy, the Milky Way, takes around 300 million years to do one complete revolution.

At the moment, our Sun is a little way above the central plane of the galaxy. One theory is that once every half-cycle, every 150 million years, our Sun dives through the central plane of the galaxy and goes through dust clouds that influence the amount of sunlight falling on our planet.

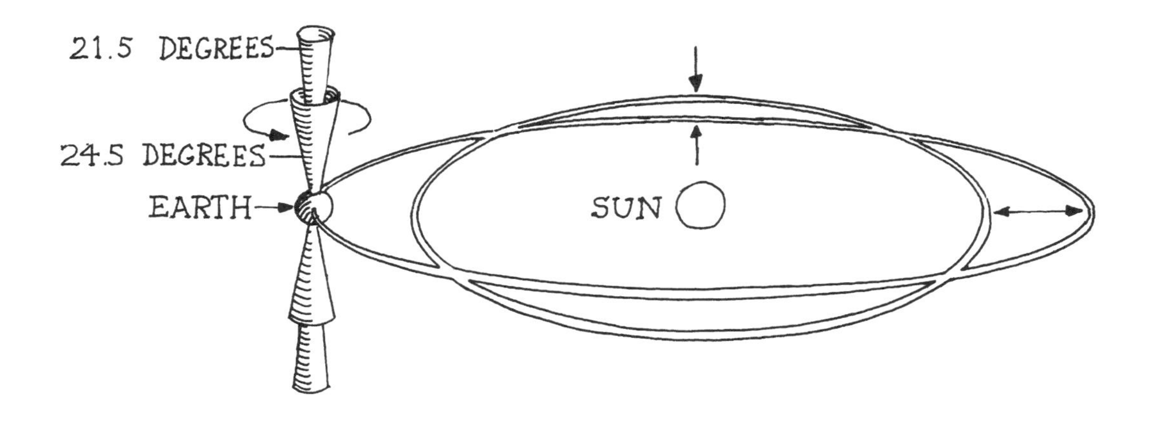

identified three factors that control the amount of the Sun's heat that lands on the Earth.

The first factor is the tilt of the spin axis of the Earth, relative to the plane of the planets. At the moment, the tilt of the Earth is around 23.5 degrees off the vertical. Over a time of 41,000 years, it varies between 21.5 and 24.5 degrees.

The second factor is the direction in which the spin axis of the Earth points. At the moment the spin axis of the Earth is pointed at the Pole Star in the Northern Hemisphere. It gradually spins around, like a very slow-moving top, before it finally comes back to point at the Pole Star again. It takes around 23,000 years to do this.

And the third factor is that the Earth's orbit varies between being almost perfectly circular to being slightly egg-shaped, and it takes about 100,000 years to do this — which fits in very nicely with the 100,000-year cycle of the advance of the ice sheets.

Since Milankovitch put forward his theory, another factor has been discovered — the plane of the Earth's orbit is not exactly lined up with the average plane of the planets, but is slightly inclined to it. It lines up with the plane of the planets every 100,000 years. This coincides with the fact that layers of cosmic dust in the ocean sediments get thicker every 100,000 years. This dust could reduce the amount of sunlight falling on the Earth.

However, these causes of the Ice Ages are just what we believe in the late 1990s, and new information could give us new theories.

The Little Ice Ages come in clusters, which happen roughly every 150 million years (I call these clusters a Big Ice Age). In each of the past clusters that we have measured, there were between 10 and several hundred Ice Ages.

Right now, we're in one of those 150-million-year clusters. The last Little Ice Age finished about 15,000 years ago, so at the moment, we're in a non-Ice Age period — also called an 'interglacial'. There most probably will be another Little Ice Age in a few thousand years, but with our current knowledge we can't say if this is true or not.

In a Little Ice Age, the ice advances from the North and South Poles and heads for the Equator. In the Southern Hemisphere, the Antarctic ice reaches to within 1,000 kilometres of Tasmania. Much of Tasmania and the more southern parts of the Great Dividing Range get covered in ice and snow. In the Northern Hemisphere, the ice is about 3 kilometres thick in North America and 2.5 kilometres thick in Europe.

And then, after roughly 100,000 years of being gripped by a Little Ice Age, the world heats up a little and the ice sheets melt and retreat.

Heinrich Events

During each Little Ice Age, and during each non-Ice Age (or interglacial), the climate can oscillate wildly.

For example, in the last interglacial, which happened about 120,000 years ago, the average temperature varied by 12 Celsius degrees and the climate flipped from one state to another in as little as 10 years!

And during the last Little Ice Age, which lasted for about 100,000 years, the climate suddenly flipped on at least six occasions — and on each occasion, enormous armadas of icebergs broke loose from the east coast of Canada, and sailed off into the Atlantic Ocean.

These events are called Heinrich Events, after Helmut Heinrich. He was looking at samples of the ocean floor from the Atlantic Ocean. To his surprise, he found rocks from Canada. The rocks and soil had travelled several thousand kilometres and been dumped at six very clearly defined points in time — ranging from 14,000 years ago to 66,000 years ago.

Immediately after each Heinrich Event, the whole planet warmed quite suddenly, and then, very gradually and slowly, began to cool down.

As the temperature dropped, the ice sheets gradually grew up to 3 kilometres thick. And then great slabs of ice, tens of kilometres long, slipped off the Canadian coast into the Atlantic Ocean. At their bottoms, the icebergs had grabbed some Canadian rocks and soil. As the icebergs melted, they dumped this debris in the Atlantic Ocean. The layers of these Canadian rocks and soil vary from half a metre thick near the middle of the Atlantic Ocean to a few centimetres thick over on the European side of the Atlantic Ocean.

Quite suddenly, within 10 to 50 years of a Heinrich Event, the local climate warmed up between 5 and 10 degrees Celsius and stayed that way for a few hundred years. And then over the next 5,000 to 10,000 years it gradually got colder again until another armada of icebergs (another Heinrich Event) sailed off into the Atlantic, and once again the temperatures flipped into a warmer state.

But the strange thing is this — Heinrich Events were not just a localised occurrence, confined only to the North Atlantic Ocean. During the 100,000 years of the last Little Ice Age, as the temperature cooled and warmed in the North Atlantic Ocean, it did the same in the glaciers of the high Andes in Chile and Peru, and in the glaciers of New Zealand.

And this is one of the great mysteries of the iceberg armadas. Did a change in the temperature of the Earth set these fleets of icebergs sailing off into the Atlantic Ocean, or did these armadas of icebergs actually cause a change in the worldwide climate? In the late 1990s, we simply don't know the answer.

Three and a half flips for the climate

In the last Little Ice Age, the ice sheets reached their peak about 18,000 years ago and then retreated. The ice sheets haven't come back, and the climate has been mostly warm-ish — but even so, in that 18,000 years the climate has done three and a half flips, from cool to warm and back again. These flips might have something to do with our planet's greatest surface heat-engine — which our ocean scientists are finally learning about.

In the first flip, about 14,000 years ago, the planet warmed up briefly. Trees grew all across Europe. Then, and we still don't know why, the Earth cooled down again.

About 12,500 years ago, the climate flipped suddenly (in less than 100 years) into a much cooler state and stayed there for 1,500 years. Most of the trees in Europe died and were replaced by Arctic shrubs and tundra flowers.

Then the climate did its second flip. Quite suddenly, in less than 100 years, the planet began to warm up. By 6,000 years ago the temperature was much like today. 2,000 years ago, around the birth of Christ, it was actually quite a bit warmer than today — so perhaps Joseph and his pregnant wife, Mary, didn't have to tramp through snow until they reached the inn. From 2,000 years ago to about 1,000 years ago, the world cooled down again.

But around AD 1000, the climate had its third flip. There was a generalised worldwide warming, which lasted until AD 1300. It was so warm that, in California and Patagonia, there were droughts for three centuries. As a result of this worldwide warming, the Vikings were able to set up successful colonies in Iceland, Greenland, Newfoundland and North America (where, if the Vikings had stayed, today's Americans would be speaking Swedish). But the climate flipped into a cold snap and the Vikings were booted out of their colonies.

From AD 1300 to about AD 1700, the world had a mini-Ice Age. All across Europe, people died in various severe winters, and in some years it was so cold that the River Thames in England froze and people were able to skate upon it.

But then, around 1850, the climate began to warm, and it's still warming today.

Giant conveyor belt of water

You can see that the climate has changed quite a bit in the last 18,000 years. What could cause such huge changes?

One candidate is a giant conveyor belt of water. This invisible conveyor belt carries chunks of water, the size of continents, from one part of our planet to another. In the Atlantic Ocean, it carries warm water from the Equator up past the coast of Europe. It carries a billion megawatts of heat! That's 30 per cent of the heat that the Sun dumps on the North Atlantic! This heat gives Spain an almost-tropical climate. This is why, when you compare two cities at the same latitude like Bödö in Norway and Nome in Alaska, you'll find that the European town is 4 degrees Celsius warmer than Nome in summer and 13 degrees Celsius warmer in winter.

As the water gives off its heat, it cools down and becomes denser. The dense water sinks and falls down in a giant vertical elevator towards the ocean floor. It then flows South, past New York, and towards the Equator. It continues past South America, and then swings left past the bottom of Africa and heads for Australia. It flows past the southern coast of Australia, sweeps around New Zealand and then heads up toward Alaska.

This giant conveyor belt carries about 100 times more water than all of the rivers in the world. At the moment, we know very little about this giant conveyor belt that carries heat and water. But we do know that it can change its path very rapidly.

The Weather Bureau can give you a 70 per cent accurate forecast for the next few days, but whether in 1,000 years time the Earth will be in an ice box or a sauna — well, nobody knows that!

Tibetan ice-block effect

Tibet is a spiritual place. It sits on the roof of the world, on the 5-kilometre-high Tibetan plateau. Some researchers now believe that this plateau may have influenced the entire planet by cooling the climate and even affecting the evolution of the human brain.

The climate of the world has been fairly predictable over most of the last few hundred million years. Until recently, it was warm and wet, like the tropics. The dinosaurs, who lived from 200 million to 65 million years ago, enjoyed a temperature about 8–11 Celsius degrees warmer, and swam in seas about half a metre higher, than we do today.

But all this changed about 50 million years ago, when India collided with Asia at the frightening speed of 20 centimetres per year (roughly four times faster than your fingernails grow). As a result of this slow but gigantic collision, the Himalayas and Tibet gradually rose as India ploughed northwards another 2,000 kilometres.

During this enormous collision, the Antarctic began to ice up and the world cooled down. On average, the world's temperatures kept on dropping, until about 2 million to 3 million years ago. It was around then that our human brain began to double in size, from 600 millilitres to about 1,200 millilitres. It could be coincidence, but Big Brains do need a lot of cooling.

One theory claims that the Tibetan plateau is responsible for cooling the world by taking carbon dioxide out of the air. We're certain the Tibetan plateau causes the annual Asian monsoons. Eight huge rivers, which include the Ganges, Mekong, Indus and the Yangtze, drain from the Tibetan plateau and its approaches. These rivers drain a total area of less than 5 per cent of our Earth's land area, but they dump 25 per cent of the minerals that reach the ocean.

The very heavy rains, combined with enormously steep slopes, cause huge erosion. The carbon dioxide that is dissolved in the raindrops forms a weak acid — carbonic acid. This carbonic acid combines with granite and limestone to make minerals which wash out and get dumped on the ocean floor. These minerals are very rich in carbon.

In fact, according to this theory, the Tibetan plateau is really a huge pump that takes carbon dioxide out of the atmosphere and deposits it on the ocean floor where it is locked away for millions of year.

The Tibet theory was created by oceanographer Maureen Raymo from Massachusetts Institute of Technology and her colleague Bill Ruddiman, a palaeo-climatologist of the University of Virginia. They claim that chemical reactions caused by the Tibetan plateau have removed so much carbon dioxide from the atmosphere that the temperature has dropped — not the Greenhouse Effect but the Tibetan Ice-block Effect.

Now the theory is in its early days, and it's not rock solid, but we do know that after about 50 million years of a steady downward drop in temperature, the Earth's climate seems to have stabilised into an oscillating series of Ice Ages and non-Ice Ages. So maybe Tibet not only chilled out the world, it also gave us swollen heads as well.

Iron in the ocean

At the moment, the human race is dumping around 6 billion tonnes of carbon into the atmosphere each year, mostly in the form of carbon dioxide. Practically all the scientists think that this carbon dioxide will set off a Greenhouse Effect, which will make the average temperature of the world go up.

In 1987, an ocean scientist suggested that by spraying finely powdered iron over the oceans of the world, we could suck this carbon back out of the atmosphere. It sounds wild, but by 1996 we'd done the experiment and he could be right.

It's all to do with 'phytoplankton'. 'Phyto' means plant, and 'plankton' comes from the Greek word *planktos*, which means 'wanderer'. So phytoplankton are tiny little plants that float through the oceans.

Phytoplankton have two major jobs in life. Their first job is to be eaten by bigger creatures. So phytoplankton are right at the bottom of the food chain. Their second job is to do photosynthesis, and turn sunlight

THE LUNGS OF THE WORLD

Many people say that the Amazon Forest is 'the lungs of our planet'. They're completely wrong. In fact, the Amazon Forest uses up more oxygen than it makes, because wood uses up a huge amount of oxygen as it decays. Oxygen mostly comes from the phytoplankton in the oceans.

into energy. At the same time, phytoplankton suck in water and split it into hydrogen and oxygen. They then combine the hydrogen and oxygen with carbon to build more phytoplankton.

Phytoplankton are really small, but they are actually the most abundant type of life living in the oceans, from the point of view of sheer weight. But there has always been one thing about phytoplankton that has puzzled the oceanographers. You can have an area of the ocean that is loaded with all of the phosphate and nitrates that the phytoplankton need for growth, but they don't grow. Instead, the area is a virtual desert. What's going on?

The late John Martin had a theory. Martin, who was an oceanographer, reckoned that the reason that the phytoplankton weren't growing, even though there was all this food around, was because they were short of one single essential nutrient — iron.

He did some measurements, and he found those parts of the ocean which were low in iron had very low growths of phytoplankton, and those that were rich in iron had very high growths of phytoplankton.

In 1994, John Martin did his first experiment on a big scale and dumped iron into the ocean. His oceanographic research ship steamed slowly backwards and

IT'S HARD TO MEASURE IRON

One problem that John Martin had was that it's incredibly difficult to get a *clean* measurement of how much iron there is in the ocean. Dust blows everywhere, and 5 per cent of the weight of dust is iron. You might sample the ocean water in clean plastic bottles straight from the factory — but plastic bottles are often made on a stainless-steel former, and some of the iron in the stainless steel goes into the plastic. And of course, the rope used to pull the bottle out of the ocean is often a steel rope — and again, the iron in the steel rope can contaminate the sample.

John Martin went to an incredible amount of trouble to make sure that he avoided iron contamination. He was soon able to measure consistently and accurately the background levels of iron in different parts of the ocean.

For example, the Antarctic is incredibly rich in all of the nitrates and phosphates that phytoplankton need to grow, and yet blooms of phytoplankton hardly ever happen. Sure enough, John Martin found that levels of iron in the ocean around the Antarctic were too low for the phytoplankton to grow.

It turns out that phytoplankton need incredibly small amounts of iron to grow. The figures are a little bit rough, but it seems that phytoplankton need one atom of iron for every 100 atoms of phosphorous, every 1,500 atoms of nitrogen and 10,000 atoms of carbon. So you don't need much iron to give the phytoplankton a kick-start.

forwards across 64 square kilometres of ocean, feeding 480 kilograms of finely powdered iron into the propeller wash. The phytoplankton grew like crazy. But within a few days, the effect wore off.

In 1996, after John Martin's death, his colleagues continued his work. This time they added the same amount of iron to roughly the same area of ocean — about 72 square kilometres. But instead of adding the iron in one hit, they added it in three doses, spaced over a week. This time, they found that the iron tended to remain at the surface. In the iron-enriched water, the phytoplankton grew like crazy and sucked carbon dioxide out of the water. This carbon dioxide originally came from the atmosphere.

Looking at the numbers from the experiments so far, it seems that we can suck one year's worth of carbon out of the atmosphere by adding one million tonnes of iron to the oceans — a few supertankers' worth.

It seems utter madness to go ahead and do a wild and unproven experiment (such as dumping huge amounts of iron into the ocean) to suck out carbon from the atmosphere. We have no idea what other unwanted effects might result. But on the other hand, since the Industrial Revolution began, we have already dumped some 200 billion extra tonnes of carbon into the atmosphere.

Are we game to walk the plank, on the basis of just a few experiments, to fight the good fight by calling up a new Iron Age to fight the pollution of the Industrial Age?

REFERENCES

Bond, Gerard C. & Lotti, Rusty, 'Iceberg discharges into the North Atlantic on millennial time scales during the last glaciation', *Science*, vol. 267, 17 February 1995, pp. 1005–1010.

Broecker, Wallace S., 'Massive iceberg discharges as triggers for global climate change', *Nature*, vol. 372, 1 December 1994, pp. 421–424.

Broecker, Wallace S., 'Chaotic climate', *Scientific American*, November 1995, pp. 44–50.

Broecker, Wallace S. & Denton, George H., 'What drives glacial cycles?', *Scientific American*, January 1990, pp. 42–50.

Coale, Kenneth H., *et al.*, 'A massive phytoplankton bloom induced by an ecosystem-scale iron fertilisation experiment in the Equatorial Pacific Ocean', *Nature*, vol. 383, 10 October 1996, pp. 495–501.

Farley, K.A. & Patterson, D.B., 'A 100-kyr periodicity in the flux of extraterrestrial 3He to the sea floor', *Nature*, vol. 378, 7 December 1995, pp. 600–603.

Kunzig, Robert, 'Earth on ice', *Discover*, April 1991, pp. 54–61.

Paterson, David, 'Did Tibet cool the world?', *New Scientist*, no. 1880, 3 July 1993, pp. 29–33.

Thomson, David J., 'The seasons, global temperature, and precession', *Science*, vol. 268, 7 April 1995, pp. 59–68.

Van Scoy, Kim & Coale, Kenneth, 'Pumping iron in the Pacific', *New Scientist*, no. 1954, 3 December 1994, pp. 32–35.

Waters, Tom, 'Roof of the world', *Earth*, July 1993, pp. 26–35.

Waters, Tom, 'Did the Himalaya warm the world?', *Earth*, January 1994, p. 10.

HOARDED HEARTBEATS & BABY-CUDDLING

When people talk about love and strong emotions, they often mention the heart. They might say, 'Absence makes the heart grow fonder', or 'I am heartbroken', or 'I love you with all my heart'. The heart is a powerful symbol and it's on the left side of your body — but does this have anything to do with why we hold babies on the left side?

The heart does more than fool with your emotions — it's an extremely efficient and long-lasting pump. But can you believe the old story that this pump has a certain number of heartbeats programmed into it, and then it stops?

Heartbeats

A few thousand years ago, Aristotle, the famous Greek philosopher and scientist, claimed that the heart was the seat of the intellect. It sounded like a reasonable theory at the time. After all, didn't the voice come out of the chest, so shouldn't the controller of that voice also be in the chest? Well, that was then, but now there's a new theory about the heart — the Piggy Bank Theory. It says that we humans are given about 2.8 billion heartbeats, and once we use them up, we die. With this theory, and a bit of clever mathematics, you can probably work the system to get an extra quarter century of life.

Back in the old days, people really didn't know exactly what the heart did. One of the anonymous followers of the Ancient Greek physician Hippocrates claimed that 'man's intelligence, the principle which rules over the rest of the soul, is situated in the left chamber'. He or she was talking about the left ventricle of the heart, which pumps red blood loaded with oxygen, at a high pressure, out into the various organs and systems of the body. Other scientists and philosophers thought that the heart controlled not intellect but strong emotions, such as love and sorrow, joy and bravery.

It took William Harvey, in 1628, to show, with a few elegant experiments, that the heart was actually a pump. But what a pump it is! This tireless organ runs for about 75 years, and roughly once each second it pushes blood through some 100,000 kilometres of blood vessels (about three times the distance around the Equator). And in an average human life, it will pump about 200,000 tonnes of blood — roughly the weight of three very large and fully loaded nuclear-powered aircraft carriers. And to shift all that blood, it will pump about 2.8 billion times.

SLOW HEART

When I used to do a lot of running, one of my fellow runners was a medical doctor. My heart rate was around 55, thanks to all the running, but my friend was a much more serious runner, with a better-conditioned heart, so he had a heart rate around 45.

Now to understand this little story, you need to know two things. First, a slow heart rate can be a sign of disease of the heart. Second, when you lose a lot of blood, your heart still tries to keep the same output of about 5 litres per minute by beating faster.

My medical friend was booked into a hospital for some minor abdominal surgery, but the doctor who examined him refused to admit him on the grounds that his heart rate was too slow! It took quite a few phone calls before the admitting doctor was convinced that this extremely low heart rate was normal for my friend.

The actual operation went well, but half a day later my friend began to suffer some internal bleeding. He knew that this was happening to him, not because of any internal symptoms but because he felt a little faint (because of loss of blood) and because his heart rate had suddenly accelerated to 60. He asked the nurse to call a doctor to examine him. But she measured his heart rate and said that it was normal at 60, and therefore there was no point in calling a doctor. Once again, it took a whole lot of phone calls before anything was done!

This leads us to our first problem. Practically all mammals, regardless of how big or small they are, seem to get allocated about one-third as many heartbeats as we get. They get only about 800 million heartbeats. Looking at the animals, it seems to be a general rule that the bigger the animal, the slower the heart rate. The other general rule is that the bigger the animal, the longer the lifespan. So a tiny mouse will use up its non-renewable quota of heartbeats in about two years, while an elephant spreads them out over half a century. If we got only 800 million heartbeats, we humans should live for just 26 years. We really have no idea why we humans get so many extra heartbeats, as compared to the other mammals.

So humans are already on top of the system, but Drs Stoller, Adler and Holland from Cleveland, Ohio, reckon we can work the system even more, and they claim to have done the numbers. Some people say that 'runners live five years longer, but they spend all of those five years running', but the good doctors disagree.

Your average human will have a resting heart rate of about 72 beats in each minute and live for about 75 years, and in that time, use up about 2.8 billion heartbeats.

But the American doctors looked at someone who, at the age of 20, decides to get fit by exercising for 45 minutes a day for 5 days a week. For that 45-minute window, you're using up more heartbeats — in fact, your heart rate rockets up from 72 to 180 beats each minute. But that short amount of exercise soon conditions your heart and

HEART-ATTACK RISK

If you have your weekends off, Monday is risky. A German study looked at over 2,500 victims of heart attack. It found that retired people, or those who do not work on Mondays, did not have any increased risk of heart disease on Mondays. But the average working population who had Saturday and Sunday off had a 33 per cent increased risk of heart attack on Mondays as compared to any other day of the week.

Big holidays are risky as well. An American study found other risk factors when it looked at 120,000 heart attacks that happened over a five-year period and which were treated at 90 different hospitals. Men had a 16 per cent higher risk of heart attack on New Year's Day. This jumped to 21 per cent on their birthdays, and to 28 per cent after Easter.

It seems like the morning after the big night before is a risky time for your heart.

makes it pump more blood on each beat, so at rest your heart can actually slow down. This new slower heart rate for most of the day more than compensates for the 45 minutes of increased heart rate.

Prime athletes can get their resting heart rates down to 30 beats per minute, but our American doctors were conservative and assumed that the heart slowed down to 55 beats per minute, from its previous 72 per minute.

So by the end of the week, you've actually used up less of those precious, non-renewable heartbeats from your Piggy Bank. And by the time you've reached 60 years of age, you've still got about 1.5 billion heartbeats left in your Piggy Bank. This works out to an extra 25 years of life, and living to around the age of 85. You could even live to 100 and have an even chance of getting that congratulatory telegram from the Queen. To get that extra quarter century of life, you'd have to exercise (from age 20 to age 60) for a total of less than one year — 332 days, to be exact.

Nobody has actually proved this theory about stretching out your heartbeats over more years. But millions of citizens around the world are doing the experiment right now, and after the next 30 years or so, the results should be in. But, on the face of it, it does seems good odds — better than 25 to 1. So ask yourself, do you feel lucky?

They used to say 'live fast, and die young', but maybe they should say 'run fast, and die old'.

Baby-cuddling

Everybody has seen a baby, and most of us have had a cuddle of one. It's easy to do — babies smell and feel so good. But there's something that has always worried

EXERCISE - GOOD OR BAD?

There is no doubt that regular exercise is good for your heart. Regular exercise, four or five times a week, will condition your heart.

But can hard physical work give you a heart attack?

If you do lots of exercise anyway, there's no increase in risk. So, when you do any hard physical work, your heart rate and blood pressure will increase to a lesser extent than if you had previously done no exercise at all.

But if you're a bit of a slob, the risk of a heart attack because of hard work is somewhere between seven and 100 times greater than if you're just sitting around the house.

So if you're not very fit, and you want to get fit, you should do it gradually. And it's probably better to do it in the afternoons. The platelets (tiny cells in your blood that are involved in clotting) are more 'sticky' in the mornings than in the afternoons — which is why most heart attacks happen before midday.

So get fit gradually, and try to do your exercise in the afternoons.

psychologists — for some unknown reason, people mostly seem to cradle a baby in their left arm.

This tradition goes back a long way. The Talmud, the ancient Jewish religious text, advises that 'A woman who begins to nurse her son should start on the left side, as the source of all understanding is from the left side.' If you look at paintings and sculptures that show an adult holding a baby, 80 per cent of the adults hold the babies on their left. Many of the two-legged animals, such as chimpanzees and orang-utans, also hold their babies on the left side.

Psychologists have been curious about this for some time. As long ago as 1835, the physiologist Ernst Weber suggested that perhaps the left side of the body was more sensitive to touch, which is why mothers liked to carry their babies there.

Lee Salk, a clinical professor in paediatrics at the Cornell University Medical School in New York, did some work in this field in the early 1970s. He worked with mothers who had just recently given birth. When he held their babies and presented the babies to the mothers directly along their centre-line, 80 per cent of the mothers would automatically hold their baby on their left side.

When he asked the right-handed mothers why they used their left hand to cuddle the baby, they said that this would leave their stronger right hand free for other tasks. The left-handed mothers said that it was natural to use their stronger, left hand to hold the baby.

Over the years, there have been all kinds of explanations proposed for why it is that about 80 per cent of men and women hold babies on their left. Lee Salk claimed that because the mother's heart is usually on the left, the baby could somehow hear the mother's heartbeat more easily and would calm down more quickly. Unfortunately, it's pretty hard to hear a heart at the best of times, and the valves, which actually make the noise, are very close to dead centre.

One interesting experiment was done with pillows that were the same size as a baby. When women were given these pillows to hold, they would hold them equally on the left or the right. But when they were asked to imagine that they were holding a baby, they would hold the pillow on the left. In another experiment, Salk looked at mothers who had given birth to premature babies and who had not been able to cuddle their babies for at least the first 24 hours. These women had no preference, and were equally likely to carry the baby on the right or the left.

People usually speak of there being just one single brain in a skull. But you actually have two quite separate brains (each with about 7 billion nerve fibres). These are joined by a tiny nerve bundle (which carries only 1 million nerves), so there's not very good communication between the two separate brains, which are sometimes called hemispheres.

The right brain is better at processing facial expressions and emotional gestures. If you are looking at someone on your left side, such as a tiny baby that you are cuddling on your left, the information from both eyes is processed in your right brain. Presumably, this will help you to bond better with someone who robs you of your sleep and who will cost you over $100,000 by the time he or she is 18 years old.

There's a similar situation happening with human ears. Research has shown that the left ear is better at recognising music and language and their emotional effects. Once again, holding the baby on the left would give all kinds of advantages to both the parent and the baby in getting to know each other better.

So while we don't fully know why we prefer to cuddle a baby on the left, it does seem that to be a well-balanced parent you need to carry a baby in an unbalanced fashion!

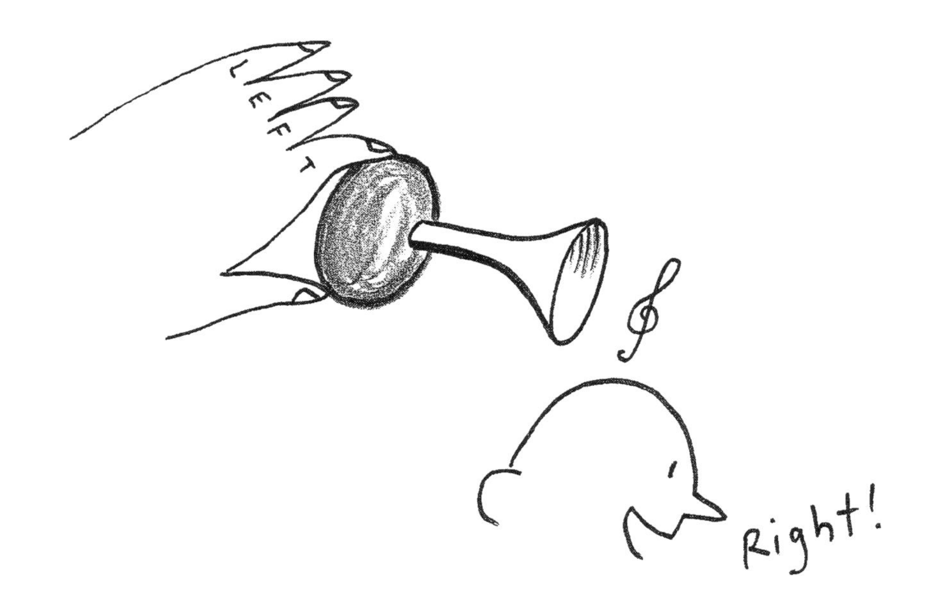

REFERENCES

Curfman, Gregory D., 'Is exercise beneficial — or hazardous — to your heart?', *New England Journal of Medicine*, vol. 329, no. 23, 2 December 1993, pp. 1730–1731.

Eastwood, Robin, 'Right versus left side', *The Lancet*, vol. 348, 5 October 1996, p. 970.

Mittleman, Murray A., *et al.*, 'Triggering of acute myocardial infarction by heavy physical exertion', *New England Journal of Medicine*, vol. 329, no. 23, 2 December 1993, pp. 1677–1683.

Salk, Lee, 'The role of the heartbeat in the relations between mother and infant', *Scientific American*, May 1973, pp. 24–29.

Sieratzki, J.S. & Woll, B., 'Why do mothers cradle babies on their left?', *The Lancet*, vol. 347, 22 June 1996, pp. 1746–1748.

Stoller, James K., Adler, Dale S. & Holland, Joe, 'Can you "run your heart out"?', *New England Journal of Medicine*, 17 March 1988, pp. 708–709.

EAT LESS, LIVE LONGER — & DRINK BEER!

From the moment we're born, we begin to die. But death is actually a fairly new invention. Death has been a natural part of life for only a small percentage of the time that life has been on our planet!

The invention of death

Our planet is about 4.6 billion years old. Life began about 3.8 billion years ago — and for most of that 3.8 billion years, living creatures were effectively immortal.

From 3.8 billion years ago to 1 billion years ago, each living creature was made up of a single cell. A single-celled creature would grow to a certain size and then split into two, and each of those two cells would then keep on growing until they divided and so forth. These single-celled creatures would never die of old age. They might die from being hit by a rock, or being run over by a fast-moving lava flow, or by being eaten by another bigger single-celled creature, but apart from that they would not die.

But about 1 billion years ago, some of the single-celled creatures evolved into creatures with many different types of cells. These creatures had some cells for thinking, other cells for moving, other cells for digesting, and so on.

And at the same time, death was invented. It seems odd, but there actually are a few advantages to dying — at least, as far as the species is concerned.

Firstly, an immortal species can't adapt to any changes in the environment — only their children, with a slightly different DNA, could. Another disadvantage of immortality is that the parents and their offspring would be fighting for the same amount of food. And thirdly, the DNA of these immortal creatures would constantly be damaged by radiation from space and by chemicals in the environment, leading to more defects in the eggs carried by the mother.

It might sound unbelievable, but we really don't know why we age. However, we do have a lot of different theories.

The first one is the environmental theory. It looks at the environment in which the cells inside living creatures survive. For example, various chemicals may gradually build up in various parts of the body, giving the cells a less pleasant environment in which to survive.

The metabolic theory of ageing is like a 'wear and tear' theory. It basically says that we have a limited number of days of life, and the harder we work the sooner we use up those days.

The autoimmune theory says that when you're born, every cell in you is immunologically identical to every other cell in your body. But as time goes by, the immunological signature of some of the cells changes and so your body's immune system starts attacking your own cells.

Another theory is the error theory. It says that as we age, we get more errors in the DNA as it divides and divides with each successive generation of cells. For example, we humans have 46 chromosomes in our cells. But in human females, the percentage of cells with the wrong number of chromosomes increases from 3 per cent in females aged 10 years to 13 per cent in females aged 70 years.

And yet another theory says that we are just plain 'programmed' to die. There's a little genetic clock inside each cell, and once it has gone through a certain number of divisions, it tells the cell to die.

As we get older, our bodies change. Our eyes can't focus as well. Our skin becomes less elastic, and our bones gradually lose calcium. Our lungs can't hold as much air as they used to. As we get older, our kidneys can't concentrate our urine so well. This means that to dump out the same amount of waste we need to add a greater volume of water.

Perhaps there is one surprisingly easy way to live for a longer period of time, in a greater state of health — and it's a simple secret — just eat less!

Eating less

In the 1930s, Clive M. McCay and his colleagues at Cornell University placed rats on a very low calorie diet. They were astonished to find that they could increase the maximum lifespan of their rats by one-third — from three years to four years. They also found that those rats on the low calorie diet stayed younger for longer, and had fewer of the diseases that the normally fed rat friends got as they grew older.

Since then, various scientists have experimented with all kinds of creatures (spiders, water fleas, guppy fish, several different parasites, and so on), and have been able to increase both the average lifespan and the maximum lifespan.

Now, there's a big difference between increasing the *average* lifespan and the *maximum* lifespan.

Increasing the *average* lifespan just means stopping unnecessary and premature deaths — such as deaths from car accidents, deaths from infectious diseases which would be easily stopped by vaccination, and deaths from heart disease.

Increasing the *maximum* lifespan is a whole different kettle of fish. If you can increase the maximum lifespan of various members of a species, you're somehow fooling around with some basic ageing process. And yet, that's what reducing the calories seems to do.

Now we know that this reduced-calorie-intake increased-lifespan thing works in quite a few animals, but we haven't fully tested it on the two-legged animals. That began only about 10 years ago, when two separate groups of scientists began testing rhesus monkeys.

It's early days yet, but already there's quite a few interesting changes. As you would expect, the monkeys that have been eating 30 per cent fewer calories weigh about 30 per cent less, and have only 10 per cent of their body weight as fat — instead of the normal 30 per cent. What is interesting is that all of their 'measures of health', even a few years into the experiment, are those of much younger monkeys. So the skinny monkeys have lower blood pressure, lower blood glucose level, lower insulin level, and much lower blood cholesterol. We have not yet done the same experiment with humans. There's no point in looking at people that have been forced by poverty to live on very low calories, because usually their low calorie diets don't give them enough essential nutrients.

If you want to increase your life expectancy, you can't begin this calorie-restriction thing when you're a child because children need lots of many different nutrients. They also burn up a lot of energy — a toddler can have four times the metabolic rate of an adult. Beginning at the age of 20 would probably give the greatest extension of life, but it's never too late to start. And it also seems that you can't simply go cold turkey and suddenly reduce your calories by 30 per cent — you need a gradual tapering off. And most importantly, with a low calorie diet you still need to have your essential nutrients. You would have to select your foods extremely carefully, and probably have to take vitamin supplements as well.

But there would be side effects, besides being hungry a lot of the time. A woman whose weight is very low might not ovulate, and so could not have babies. And this lack of ovulation could lead to increased osteoporosis.

Indeed, Professor Stewart Truswell, of the Sydney University Human Nutrition Unit, said that while these scientific studies were interesting, 'it certainly wouldn't make me go and eat less than I normally eat'.

On the other hand, most of us could certainly afford to lose a little weight, and if we did it in a gentle fashion there would be hardly any side effects at all. But maybe one group of American scientists has the right commercial idea. They're working on a drug that will give you the benefits of not eating much — not only will you be able to have your cake and eat it, but you can eat your way to good health!

Beer belly

Beer is a part of the Australian way of life. Australia is one of the top three beer-consuming countries, the others being Germany and Czechoslovakia. Australia's average beer consumption is around 150

LIGHT BEER SAVES LIVES

Alcohol is one of the big killers in our society. According to David Crosbie, the Chief Executive Officer of Alcohol and Other Drugs Council of Australia: 'Drunkenness is a major public health issue, accounting for 34 per cent of drownings, 34 per cent of all falls, up to 20 per cent of suicides, up to 30 per cent of domestic violence incidents, conservatively, more than 50 per cent of all public disorder crimes.'

Crosbie said that, in Australia, more than 10 people die every day because of the misuse of alcohol, and most of these people are under the age of 35. 'So it is not excessive use over a long time but *drunkenness* that is a major public health problem.'

He said that thousands of lives have been saved on the roads of Australia, because Australians are drinking more low-alcohol or light beer (under 3.5 per cent) and less normal beer (5 per cent or more). This then reduces the number of people on the roads who are driving with high blood alcohol levels. He said that in society in general, as the consumption of full-strength beer lessens, so do the health and social problems associated with alcohol.

litres per person per year. One supposed side effect of beer drinking is the beer gut — that big bulge out the front. But according to two nutritionists from Deakin University, not only does beer have fewer calories or kilojoules than skimmed milk, there's no such thing as a beer gut!

Beer has been around for many thousands of years. Some 5,500 years ago, in what is now Iran but was then Mesopotamia, people made jugs that had curious deep grooves crisscrossing the inside. Virginia Badler from the University of Toronto analysed a chemical residue in these grooves. She found it was calcium oxalate, which is today also called 'beerstone', because it's left behind from the process of brewing beer.

To make beer, you need yeast. Yeast converts simple sugars into alcohol and carbon dioxide. By trial and error, we humans settled on barley as the source of our sugars. If you slice open a barley seed, you will find inside a tiny embryo (which will turn into a new barley plant), and a huge amount of food supply in the form of complicated sugars (to feed the embryo while it's growing).

Unfortunately, there's one small problem. The complicated sugars can't be eaten by the yeast. But if you wet the barley seed with water, it will sprout tiny shoots. These tiny shoots make chemicals that will break down the complicated sugars into simple sugars. Probably our ancestors found this out by accident, when some wet barley seeds were settled upon by a yeast, which turned them into alcohol.

So, three of the ingredients of beer are barley seeds, water to turn the barley into food for the yeast, and yeast to turn them into alcohol. The fourth ingredient is hops,

BEER & SALTY POTATO CHIPS?

Why do beer and salty potato chips go with each other so well? US scientists accidentally found the answer when they were trying to answer a different question: 'Why is salt such a good flavour enhancer?'

We humans have been using salt for thousands of years in our cooking. It seems to bring out the flavours of the food. But laboratory experiments show that table salt by itself does not boost individual flavours! What's going on? It turns out that salt acts as a 'flavour filter' — it actually suppresses unpleasant tastes, like the bitter taste of beer. Once these bitter tastes are suppressed, the sweet tastes can come to the front.

The scientists tested three different liquids: a bitter-tasting chemical (urea), a sweet-tasting sugar (sucrose) and a salt (sodium acetate).

They found that the salt suppressed only a little of the sweetness of the sucrose, but that it suppressed most of the bitterness of the urea.

They then mixed the sucrose and the urea together — as expected, the volunteers said that it tasted bitter.

When the experimenters added some salt, the sucrose–urea solution tasted more sweet and less bitter. The salt was suppressing the bitterness again!

So now you know why you have a sudden need to eat salty potato chips with your beer — because the salt is not a flavour enhancer, but a 'flavour filter' that stops the bitterness coming through.

a bitter herb. It gives the beer a nice taste, and also adds some natural antiseptics to keep out unwanted yeast and bacteria. As a bonus, the hops also contributes tannins, which combine with proteins in the beer and stop them from forming an unattractive haze.

Some 5,500 years ago, the inhabitants of Mesopotamia used 40 per cent of their barley crop to make eight different beers. At 220 million tonnes per year, barley is the world's sixth largest food crop today — and practically none of it gets eaten and most of it ends up as beer.

Well, it might end up as beer, but — according to Professor Kerin O'Dea and Dr Kevin Rowley from Deakin University in Victoria — it doesn't necessarily end up as a beer gut. O'Dea and Rowley put up a bunch of arguments.

First, beer has fewer kilojoules, or calories, than skimmed milk. Second, alcoholics are usually quite lean. Third, alcohol uses up a lot of energy just to digest it. Suppose on your plate you have some food rated at 100 units of energy. You have to chew it, swallow it, mash it with your stomach and finally absorb it into your bloodstream. This digesting takes 4 units of energy for your average food, but 8 units of energy for alcohol – leaving behind less energy to turn into fat.

It seems that abdominal fatness, which we sometimes call a beer gut, isn't in fact a beer gut. It is a beer-and-salty-peanuts-to-make-you-thirsty gut, or a beer-and-salty-chips-to-make-you-thirsty gut, but it isn't a straight beer-by-itself gut.

So, don't hate beer. Beer has been important for thousands of years, and in moderate amounts will be important in the future. After all, that great musician and philosopher Frank Zappa said: 'You can't be a Real Country, unless you have a beer . . . It helps if you have . . . a football team, or some nuclear weapons, but at the very least, you need a BEER.'

REFERENCES

'Aussie beer belly exposed as a myth', *Daily Telegraph Mirror* (Sydney), 26 March 1995, p. 19.

Coghlan, Andy, 'Pint pots designed to banish bitterness', *New Scientist*, no. 1848, 21 November 1992, p. 8.

Dayton, Leigh, 'Eating less seen as key to long life', *Sydney Morning Herald*, 4 March 1997.

'Eating less could extend life', *Australian Doctor Weekly*, 14 March 1997, p. 37.

English, Ruth & Lewis, Janine, *Nutritional Values of Australian Foods*, Department of Community Services and Health (booklet), Commonwealth of Australia, 1991.

How Is It Done?, Reader's Digest (Australia), 1990, pp. 346, 394.

'Is alcohol fattening?', *Health Reader*, vol. 6, no. 9, April 1995, pp. 1–2.

Mortimer, Derek, 'Beer saves lives', *Medical Observer*, 2 May 1997, p. 28.

O'Dea, K. & Rowley, K.G., 'Treating eating problems', in I.D. Caterson (ed.), *Nutrition and Overweight*, no. 11, Servier Laboratories, Melbourne, 1994.

'Salty secrets revealed', *Sunday Mail* (Qld), 8 June 1997, p. 69.

Santos, Levin, 'Dead man's curve', *The Sciences*, January–February 1997, pp. 10–11.

Weindruch, Richard, 'Caloric restriction and ageing', *Scientific American*, January 1996, pp. 32–38.

CLAY GIVES LIFE & IS GOOD TO EAT

We humans are the two-legged animal that uses fire. We've been adding fire to clay to make pottery for over 8,500 years. We eat our food off flat circles of clay and take our drinks from open-ended cylinders of clay. Some people bypass the food and the drink, and eat clay itself — often for very good medical reasons! And many scientists think that clay was involved in the appearance of life on our planet.

Clay made life

'Where did life come from?' is another of The 10 Big Questions. One theory is that life came from a very common substance — clay!

The Earth and the Sun formed about 4.6 billion years ago. But within a surprisingly short time, only about 800 million years, there was life on our planet. We know this because scientists have seen the fossils of tiny single-celled creatures called stromatolites in some 3.8-billion-year-old rocks dug up in Western Australia.

When our Earth was very young, long before plants made oxygen, the atmosphere was rich in hydrocarbons like methane and ammonia. Back in 1952, Harold C. Urey

CLAY, ORIGIN OF LIFE

One theory about clay being the basis of life is that clay was the backbone upon which interesting chemical reactions could happen.

But there's another theory rapidly gaining popularity. It says that clay had the first self-replicating molecules. It certainly has the right chemical structure to do this.

and Stanley L. Miller put these and other gases in a small glass flask, and passed electrical sparks through the mixture to mimic the effect of bolts of lightning. To everybody's vast surprise, within a week they found the chemicals of life — carboxylic acid, aldehydes and amino acids.

If you join up a whole bunch of amino acids in a line, you end up with proteins, which are found in all living creatures. For example, insulin is a protein that is made from about 100 amino acids. Some proteins are enzymes (like in your favourite washing powder) which push chemical reactions along. But it's not that easy to turn single amino acids into proteins.

Firstly, the amino acids will join up only if they're held in exactly the right orientation and alignment. Imagine you are looking at two old-time ballroom dancers who want to do a waltz. They will fit together only if they are facing each other. Imagine they are separately whizzing across the dance floor and spinning around at the same time. The chances are pretty small that they will accidentally bump into each other at the same instant that they are facing each other.

The second problem is you have to add energy to join amino acids together to make proteins.

It turns out that an incredibly common material — clay — can solve both of these problems. Clay is what you get when rocks such as granite get attacked by the weather for a few million years. Clay is made up of tiny particles less than 4 microns across (for comparison, a human hair is about 70 microns across). These tiny particles are, in turn, made of millions of repeating units of identical crystals. These crystals are arranged in layers, which are usually electrically charged. The crystals are made from silicon, oxygen, magnesium, aluminium, iron, potassium and so on. There are about eight different types of clay, and they differ in how far apart the layers are, what particular shapes the crystals have, and what electrical charges the layers have.

One type of clay is kaolin, which is used among other things to coat paper. Kaolin accounts for 20 per cent of the weight of paper. It makes the paper more opaque, and also makes photographs possible by reducing the amount of ink the paper will absorb. But kaolin also has enormous surface area, thanks to all the layers. In a tiny cube of kaolin just 1 centimetre across, the layers have a total surface area of 2,800 square metres, which is over half the size of a cricket field!

Clay has tiny electrically charged pockets, which can grab and hold amino acids so that each amino acid is in the right alignment to join up to another amino acid. Think back to our dancers. If one of them is standing still, the other one has a better chance of joining up for a waltz.

FIRST SCULPTOR

Dibutades, also called Butades of Sicyon, who lived around 600 BC, was supposedly the first human to model in clay, according to the Roman writer Pliny the Elder. Apparently the daughter of Dibutades had fallen in love with a young man, and, seeing his shadow fall upon a wall, she quickly drew the outline of his face on that wall. When her father saw this outline, he quickly made in clay a face of this young man and baked it. His profession was to bake and sell clay tiles. He soon branched out into decorating the ends of roof tiles with human faces.

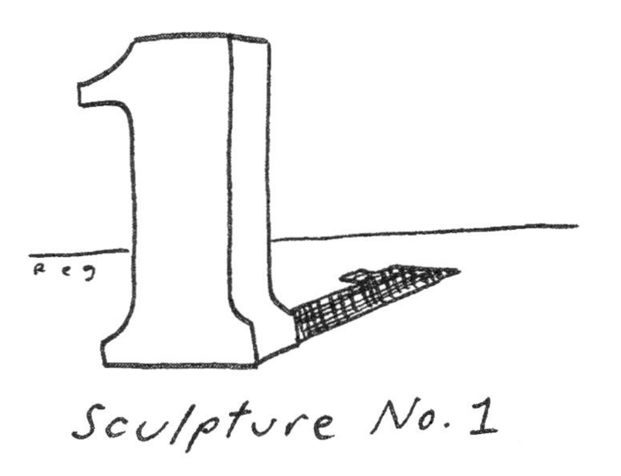

And scientists from NASA Ames Research Laboratories have found that clay can store all kinds of energy, from gamma rays to mechanical energy, and give it off later.

Maybe the prophet Job was very close to the truth when he said in the Bible (Job 33:6): 'I am just like you before God; I too have been taken from clay.'

Clay eating

Clay is an earth material that is soft and pliable when it's wet, and hard when it's dry. Clays are an essential component in the soil, and are vitally important for the growth of plants. They also have a major economic impact in our society. Clay is used in pottery, bricks, tiles, coatings on paper, medicines, wool scouring, rubber tyres, and drilling mud in drilling rigs. But clays have another, more fundamental use — many people, especially pregnant women, eat clay.

In Holmes County in Mississippi, African-American women will still meet in the cool of the evening to share a 'heaping handful' of clay — about 50 grams. Not just any dirt will do. It usually has to be dug from over 50 centimetres underground (where there are fewer parasites), or freshly exposed (after heavy rain). Sometimes it's eaten raw, and sometimes it's baked — often over a wood fire to give a 'smoked' flavour. One woman told of how her family gradually picked at their chimney, which was made from clay and straw, until the chimney began to collapse. Sometimes clay is eaten as is, by the small spoonful, and sometimes it's sprinkled over other foods, such as ice cream.

In Holmes County, the habit of eating dirt or clay is mainly practised by women (especially pregnant women), and by young children. The practice is called 'geophagy', from *geo* meaning 'earth', and *phage* meaning 'eat'. Geophagy has been practised

GREEKS ATE CLAY

When the Greeks ate their 'sealed earth', they were very fussy about what clay they used. The clay had to be collected before sunrise on 6 August, when a priestess would go into a few special caves on the island of Lemnos to get it. She would then mix it with the blood of a goat, squash it down into a small cylinder the size of a jelly bean, and mark it with the figure of a goat!

The Greek doctor Galen knew of the reputation of these medicinal clays from Lemnos, especially when they were used as an antidote to poisoning. He personally visited Lemnos and took some 20,000 lozenges back to Rome.

The clay probably worked. At that time in Rome and Greece, mercury was a very common ingredient in many poisons. This clays binds to mercury in the gut, and carries it safely out of the body before it can be absorbed into the bloodstream.

in North and South America, Africa, Northern Europe, the Mediterranean, the Pacific Islands, and in Australia. The ancient Greeks used *Terra sigillata*, or 'sealed earth', for many medical uses. The ancient Romans thought it could be an antidote to various poisons, and even today, kaolin and bentonite, both of which come from clay, are used in anti-diarrhoea medicines. This could be one of the reasons why some pregnant women eat clay — to reduce the gut side-effects of morning sickness.

There seem to be three main reasons why people eat clay — to ward off starvation, to receive extra nutrients, and to neutralise toxins.

It's well known that starving people will eat clay. Finland had a famine in 1832, and to survive the locals ate 'foodstuffs containing a meal-like silicaceous earth mixed with real flour and tree-bark', according to the American ethnographer Berthold Laufer. But this is just a case of the earth being used to bulk up real food.

Another reason that people might eat clay and dirt is because of the potential nutrients in it. These include iron, silicon, magnesium and potassium. However, some of these nutrients are not in a form that is available to the human gut. And in some cases, the iron will actually bind to the clay and be taken out of the system, leaving the clay-eater with an iron deficiency.

But there are good reasons for believing that geophagy can neutralise some toxins in food. Take the example of the colourful parrot called the macaw. Colonies of macaws that live in a certain Peruvian rain-forest will grab a clawful of clay from a local cliff. These macaws don't eat much clay when food is plentiful. But when they fall on hard times, they're reduced to eating seeds that are rich in toxins such as tannins and alkaloids, which they neutralise with a clay chaser.

In the Andes in Bolivia and Peru, the local Indians eat bitter potatoes. They first get some local clays, mix them with water to make a mush, and then dip the potatoes

EATING CLAY CAN BE BAD

Certain clays can cause iron deficiency, or anaemia. The clay binds to iron in the gut, and carries it out of the digestive system. David Livingstone, the famous explorer of Africa, blamed the eating of clay, or 'safura', for making pregnant women weak, pale and short of breath (all symptoms of iron deficiency and anaemia). In his *Last Journals*, he wrote: 'Squeeze a fingernail, and if no blood appears beneath it, safura is the cause of bloodlessness.'

One particularly nasty, and ironic, result of being iron deficient is that your gut will absorb lead better. This means that children from poorer families, who are iron deficient as a result of eating clay, are also at a very high risk of getting lead poisoning.

Another potential hazard with eating clay is that certain fine clays can stick to the lining of the gut and completely obstruct it, so you need a bit of surgery to clear your bowels again.

into the mush and immediately eat them. On the other side of the Equator, the Hopi, Zuni and Navajo Indians have the same habit. It turns out that all the potatoes that the locals eat contain a family of toxins called glycoalkaloids, and that the local clays bind to, and neutralise, these toxins.

Acorns were a popular food with the Pomo Indians of Northern California and with the natives of Sardinia, an island in the Mediterranean Sea. Acorns contain tannins, which are bitter and acid to the taste. These peoples used clay to neutralise the tannin. The clay does not bind to the tannin, but instead turns it into other, less toxic chemicals.

By neutralising toxins, clay gives us a bigger range of foods that we can eat — what you might call a well-grounded diet!

REFERENCES

Anderson, Ian, 'Does the answer lie in the soil?', *New Scientist*, no. 1481, 11 April 1985, p. 3.
Cairns-Smith, A.G., 'The first organisms', *Scientific American*, June 1985, pp. 74–82.
Dickerson, Richard E., 'Chemical evolution and the origin of life', *Scientific American*, September 1978, pp. 62–78.
Frate, Dennis A., 'Last of the earth eaters', *The Sciences*, November–December 1984, pp. 34–39.
Jones, Timothy, 'Well-grounded diet', *The Sciences*, September–October 1991, pp. 38–43.
Radetsky, Peter, 'How did life start?', *Discover*, November 1992, pp. 74–82.

I didn't used to have such hairy legs....
reg

LEECHES SUCK!

If you've ever done any bushwalking, you've probably been visited by a leech. Today it may seem strange, but in general, the medicinal value of leeches got very good press in the ancient world. Even as recently as 1850, leeches were highly prized (and priced). After a gap of a century, the medical profession has fallen in love with them again.

Specialised suckers

Leeches don't have noses, but luckily they can breathe through their skin. They have two hearts, anything between two and eight eyes, they drink your blood, and they can go hungry for a year after a good feed.

Leeches are true hermaphrodites — they're boys and girls at the same time — so inside each flat little leech body there are sperm and eggs. When two leeches decide to have some naked fun, they give each other a cuddle with their bodies lying at right angles to each other. Soon, each leech

BASIC LEECH FACTS

Leeches are flat little animals that are closely related to earthworms. They belong to the *Annelida* phylum, which is a group of animals without backbones and whose bodies are made up of many rings or segments. (*Annelid* comes from New Latin and means 'ringed ones'.)

There are about 10,000 species of annelids divided into four main classes: polychaeta ('many bristles'), such as marine worms; oligochatea ('few bristles'), such as earthworms; archiannelida, which is a fairly small group of simple marine worms; and the hirudinea — our friends, the leeches.

There are about 650 different species of leeches worldwide. They can live on land, in seawater or in freshwater. There are only 40 species of land leeches known in Australia, with probably another 100 or so yet to be discovered.

Leeches are quite hardy. Not only can they live for a whole year once they've had a good feed, they can survive a drought by letting themselves dry right out. They can even withstand a bushfire by burrowing under stones.

dumps a little capsule full of sperm on the outside of the other leech. The capsule dissolves its way into the body of the other leech, a bit like the Universal Dissolving Liquid that came out of the alien creature in the movie *Aliens*.

Some of the sperm get carried by the internal bodily fluids of the leech to the ovaries, where they pass inside and fertilise the eggs. And the hole in the body? That repairs itself in about three days. This rather odd type of sex is called 'hypodermic impregnation'.

Leeches have strong muscular suckers at each end, which they can use for moving around or for clinging to their prey. But at the mouth end, the sucker is specialised for sucking blood. The leech opens his/her little mouth, and attaches himself/herself to the intended victim. Inside the mouth are three rotary discs arranged in the shape of the letter Y — a bit like the Mercedes-Benz symbol. There are about 60 tiny teeth on each disc, and as the disc oscillates the teeth easily cut through the skin of the intended victim. Leeches suck out 10–50 grams of blood over 20 or 30 minutes, and then, full as a goog, they drop off.

To help them get away with this, they have a whole bunch of really exotic chemicals.

Firstly, there is the anaesthetic, which means that the victim doesn't feel any pain.

Another chemical is a vasodilator, which opens up the veins of their victim, allowing the blood to run freely.

There's also a strange chemical called a 'spreading factor'. It pushes apart the cells at the scene of the bite. This means that all the chemicals that the leech wants to inject can penetrate deeply into the victim. It turns out that this spreading factor actually dissolves a coating that protects bacteria. So it makes the bacteria more vulnerable and more easily attacked by the victim's immune system. This may have something to do with the old rumour that leeches can help cure some diseases.

Leeches also make anti-coagulant chemicals. When you get a cut, your blood will automatically clot to stop it leaking out all over the place. But the leeches have evolved two different chemicals — one to stop blood from clotting, and another chemical to dissolve any clots that have already formed.

There are also a few reports that, in the saliva of the leech, there is another chemical that can help slow or even stop the spread of cancer. Some mice were deliberately infected with a type of cancer — a sarcoma. A special protein from the salivary gland of the Mexican leech stopped the sarcoma cells.

The leech cannot digest our blood by itself, because it does not make its own digestive enzymes. But it has evolved a happy relationship with a bacterium, *Aeromonas hydrophila*, which lives in the leech's gut. Not only does this bacterium digest the meal of stolen blood for the leech, it even makes an antibiotic which kills other bacteria.

All these chemicals could one day be useful to us humans.

Medicinal leeches

Leeches are good blood-takers. They have their own anaesthetic, so they don't hurt the victim. And it is easy to adjust the amount of blood taken. Simply add more leeches for more blood loss, and add a little salt to make them stop. (Salt works by making the leeches vomit. Of course,

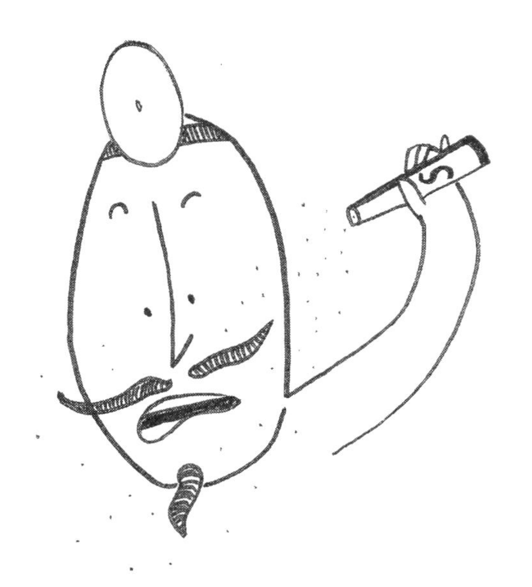

leeches vomit through their mouths, which is also what they use to grab onto you with. So if they want to vomit, they have to let go.)

In general, leeches got very good press in the ancient world. It was thought in those days that some diseases were caused by bad blood, and so leeches became popular as the most painless and hygienic way of removing this bad blood.

In the 2nd century AD, the Greek physician Galen mentioned using leeches to remove blood. Ten centuries later, another manuscript said that blood-letting 'makes the mind sincere, aids the memory, reforms the bladder, warms the marrows, opens the hearing, removes nausea, invites digestion, evokes the voice, moves the bowels, and removes anxiety'.

The *Regimen Sanitatis Salernum*, a manuscript written in the 13th century, even had a special jingle to help the blood-letters, who were mostly barbers, to remember when were the best months to use leeches:

'Three months there are in which 'tis
good to ope a vein,
Three special months, September,
April, May,
Three months when the moon bares
greatest sway.'

Leeches were popular for ages, and then — from 1820 to 1850 — a leech-mania swept over Europe.

In 1824, one single consignment of leeches from Germany to the United Kingdom contained 5 million leeches. Each leech sold for today's equivalent of about US$50. During this period of leech-mania, about 360,000 litres of blood were drained each year in France. To keep up with the

LEECHES GET BAD PRESS

The medical uses of leeches were mentioned by a Greek writer in the 2nd century BC and also by Indian writers.

Two ancient texts — Pliny's *Natural History* and the Talmud — warn about the dangers of leeches. They advise that travellers should be careful of getting nasal leeches, *Linnatis nilotica*, up their noses while drinking unfamiliar water. Even today, in Central Africa a species related to this leech occasionally causes death by suffocation.

In the 5th century BC, Herodotus described how crocodiles in the Nile would open their mouths to allow birds to wander in and out to pick leeches off their gums.

REMOVING BLOOD, WITHOUT LEECHES

There were a few different methods of drawing blood, if you didn't use leeches.

One method was 'venesection' — opening a vein with a sharp instrument.

A slightly more complicated method was 'cupping', in which a piece of burning material was placed inside a cup and then the cup was put rapidly on the skin. A weak vacuum would be created after the gases of combustion had cooled down, which would draw the blood into the cupped area. If huge blood blisters formed — and they often did — these would be cheerfully sliced and then drained of blood. There was a variation called 'wet cupping', where the barber would first slash the skin with a knife, so that more blood could flow out much more easily.

demand, France had to import some 40 million leeches every year. In Paris alone, in the years 1827 to 1836, some 6 million leeches were used every year.

LEECHES LED TO BARBERS' POLE

Thanks to the frenzy of medical blood-letting, in which the leeches played a very large part, we now have the heritage of the modern-day barbers' pole. The red stripe on the barbers' pole reminds us of the blood that they would take, while the white stripe represents the bandages that they would use to wrap the patient with afterwards. The ball on the top represents the bowl in which they would collect the blood.

In fact, a French physician called François Broussais (1772–1838) is given the credit for introducing leech-mania to France. On one occasion when he was feeling a little uneasy in his tummy, he treated himself with 15 separate applications of about 50 leeches, spread over some 18 days. He inspired a hot fashion item in the trendy salons — *Robes à la Broussais* — dresses with leeches embroidered into them!

GATHERING LEECHES

By 1802, leeches had become so popular in England, that they were now becoming scarce. Around this time, William Wordsworth wrote his poem 'Resolution and Independence', in which an old man speaks to him of gathering leeches:

> **'He with a smile did then his words repeat;**
> **And said, that, gathering leeches, far and wide**
> **He travelled; stirring thus about his feet**
> **The waters of the pools where they abide.**
> **"Once I could meet with them on every side;**
> **But they have dwindled long by slow decay;**
> **Yet still I persevere, and find them where I may."'**

Leech-mania caused such a shortage of leeches in Europe that leeches were imported from as far afield as Australia.

But in the middle to late 1800s, a new scientific fervour swept the world and leeches fell out of favour. After all, asked some doctors, why remove the blood, the very source of a patient's vitality, when a patient is at his or her weakest?

A new-found respect

Leeches became popular again in the early 1960s when John Nicholls, a Harvard neurologist, realised that they were the perfect experimental animal for him. He wanted to know how nerve cells talked to each other.

The medicinal leech, *Hirudo medicinalis*, has some of the largest and easiest-to-get-at nerve cells in the animal kingdom. They are so big, Nicholls could even put electrodes into each nerve cell to measure its electrical behaviour.

Whereas we humans have some 10 billion nerve cells, leeches have only 10,000. Whereas most of our nerve cells are all clumped together in a three-dimensional lump called the brain, the nerve cells of the leech are spread out in a thin layer, making them very easy to study.

Much of our basic knowledge of neurophysiology came from the humble medicinal leech.

By 1983, thanks to our expanding population encroaching upon its habitat, the medicinal leech (the one preferred by nine out of 10 doctors) existed in only 20 habitats in the whole world! The medicinal leech was so threatened that it was officially declared an endangered species. And it was around this time that plastic surgeons began to appreciate the medicinal leech for the superb animal that it really is.

In the early 1980s, microsurgery became very advanced. Surgeons began attempting more complicated procedures, such as reattaching the ears and fingers of small children. There was no real problem in attaching bone to bone, muscle to muscle, skin to skin, or even artery to artery. But there was real trouble in joining up the

LEECHES IN THE MOVIES

In the 1951 movie *The African Queen*, Charlie Allnutt (played by Humphrey Bogart) is not fond of leeches. As he emerges from a swamp, covered with leeches, he shouts: 'If there's anything in the world I hate, it's leeches — the filthy little devils!'

small veins that would carry the blood out of the reattached part. These tiny veins are about as thick as a human hair — and they are as strong as wet tissue paper! You just cannot sew them.

The blood could get into the attached part via the artery, but it could not get out, because there were no veins to carry it away. The ear or finger would get purple, swollen and painful, and the operation would begin to look like a failure. Western medicine had done all it could.

Enter the leech!

One or two leeches, applied every six-or-so hours, would remove the old deoxygenated blood and allow new blood full of oxygen and nutrients to flow into the reattached part. The small amount of blood that the leeches took from the patient didn't really matter, and in fact some medical studies seem to claim that blood-letting actually improves mental sharpness. And after several days, the body's own wonderful repair mechanisms would manufacture new veins. The reattached part would lose its swollen purple look and gradually become pink and warm. The leech had saved the patient's body part.

So it seems like the little suckers have given back as much as they have taken from us!

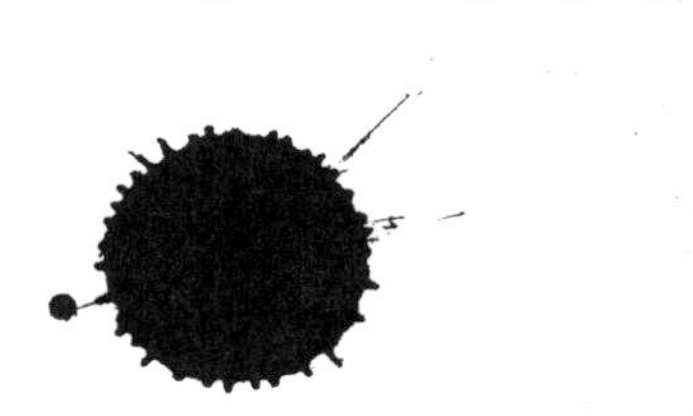

REFERENCES

Baerheim, Anders & Sandvik, Hogne, 'Effect of ale, garlic, and soured cream on the appetite of leeches', *British Medical Journal*, vol. 309, 24–31 December 1994, p. 1689.

Conniff, Richard, 'The little suckers have made a comeback', *Discover*, August 1987, pp. 85–94.

'Leeches make an anticancer protein', *New Scientist*, no. 1582, 15 October 1987, p. 31.

Lent, Charles M. & Dickson, Michael H., 'The neurobiology of feeding in leeches', *Scientific American*, June 1988, pp. 78–83.

Road, Alan & Macken, Clare, 'Welcome back, little bloodsucker', *Reader's Digest*, August 1995, pp. 161–164.

Sims, Graem, 'Leech mania', *Australian Geographic*, April–June 1989, pp. 34–47.

Tusynski, George P., Gasic, Tatiana B. & Gasic, Gabriel J., 'Isolation and characterisation of Antistasin — and inhibitor of metastasis and coagulations', *The Journal of Biological Chemistry*, 15 July 1987, pp. 9718–9723.

GRAVITY WAVES

Way back in 1687, Isaac Newton published his Laws of Gravitation — supposedly after he saw an apple fall in his orchard. Ever since, we've been able to predict very accurately what gravity *does* — just look at the success of our spacecraft in Deep Space Missions. But we still don't know exactly what gravity *is*. It's only now, over 300 years after Newton published his epic work, that physicists have finally begun to build machines to try to pick up gravitational waves.

Gravity - the dominant force shaping our Universe

Philosophers have always tried to understand how the Universe works. Aristotle, around 330 BC, said there were four basic forces or elements running the Universe — Earth, Water, Air and Fire. He said that each of these four basic elements somehow 'knew' its own 'natural' place and that it would always try to get there. So some objects would fall to the Earth, because they had a lot of 'Earth' in them. Smoke, on the other hand, would rise because it had a lot of 'Air' in it.

THE FOUR FORCES OF THE UNIVERSE

The Strong Nuclear Force holds the nucleus of the atom together. This force is carried from place to place by particles called gluons. The Strong Nuclear Force is the strongest of all the forces, but it works only over the very tiny distance of 10^{-15} metres (one-thousandth of one-millionth of one-millionth of a metre).

The next weaker force, the Electromagnetic Force, is 137 times weaker than the Strong Nuclear Force. The Electromagnetic Force is involved in keeping electrons in their orbits around the nucleus of atoms, and it's carried by particles called photons. The Electromagnetic Force has an infinite range. It's involved in radio and TV waves, radar, and beams of light.

The next force is the Weak Nuclear Force, and it controls radioactive decay and the forces that make nuclear weapons explode. It's very feeble compared to the Strong Nuclear Force, with a strength of 10^{-5}, 10^{-9}, 10^{-13} (one hundred-thousandth, one-billionth, or one-tenth of a trillionth), depending on the particles involved. The Weak Nuclear Force is carried by particles which are called the W and Z particles. It works over an even shorter range than the Strong Nuclear Force, 10^{-18} metres (one-millionth of one-millionth of one-millionth of a metre).

The last force that runs the Universe is the Gravitational Force. It controls the stars in their orbits around the centre of the galaxy, and the planets in their orbit around our star called the Sun. Like the Electromagnetic Force, the Gravitational Force reaches to the ends of the Universe. It's supposed to be carried by the graviton particle, but nobody's ever actually measured one. In fact, while a thin piece of aluminium foil is all you need to stop an electromagnetic photon of light or a radio wave, the entire mass of our galaxy with its 400,000 million stars would probably not stop a graviton!

Modern physicists agree with Aristotle that there are only four forces in the Universe — but they call these forces the *strong nuclear force*, the *electromagnetic force*, the *weak nuclear force* and the *gravitational force.*

We can, to a very small degree, manipulate the strong and weak nuclear forces with particle accelerators, nuclear reactors and hydrogen bombs. We're very good at fooling around with the electromagnetic force — we can generate electromagnetic waves with radio and TV transmitters, and we can pick them up with radio and TV receivers.

But gravity is different from the other three basic forces. The only way that we can generate a gravitational field is by piling up a whole lot of matter together in the one spot — we can't 'make' gravity with a little black box. Also, gravity affects all types of matter and energy, whereas the other three forces affect only a few types of particles. Strangely, the gravity force always attracts (it never repels), while the electromagnetic force can either attract or repel. And finally, even though the gravity force is the weakest of the four fundamental forces in the Universe, it is also the most important or dominant force shaping our Universe because its effects reach out to the edge of the Universe, and because gravity works on all types of matter.

There's another way that gravity is very different from the other three forces. Space–time has four dimensions — three space dimensions (which are left–right, back–forward and up–down) and time (which normally goes forward at the rate of one second per second). The strong and the weak nuclear forces and the electromagnetic force act with space–time as a

LOSE WEIGHT IN ALICE SPRINGS

One way to lose weight easily is to go to Alice Springs! The Earth's gravitational field is about 0.05 per cent weaker in Central Australia than it is at the coast — because the rocks there are less dense. So if you weigh 100 kilograms in Sydney, you'll weigh about 50 grams less in Alice Springs — but unfortunately, you'll still look the same.

background. But gravity is a distortion of the very fabric of space–time itself.

So if you imagine space–time as being a smooth featureless sheet, then our Sun creates a slight dip or hollow in this sheet and anything that passes near the Sun gets pulled towards the centre of this hollow.

According to Einstein's Theory of General Relativity (which is really a Theory of Gravity), an accelerating mass will give off a gravity wave which will ripple outward through the background fabric of space–time. (This is very similar to the fact that an accelerating electrical charge will give off an electromagnetic wave.)

As a gravitational wave caused by some Really Big Event in the Universe sweeps past you, it first compresses and then expands the fabric of space–time as it ripples through. This compression and expansion is very small. When the Supernova SN 1987A exploded in 1987 in a nearby galaxy called the Larger Magellanic Cloud, it caused gravitational ripples in the fabric of space–time. By the time these gravitational ripples swept past

HOW CURVED IS SPACE?

According to the physicist Bryce S. DeWitt, you can visualise the curvature of space–time just by throwing a ball into the air. If you throw a ball up to a height of 5 metres, it will be in the air for two seconds — thanks to the gravity of the Earth. In those two seconds, light will travel some 600,000 kilometres. So the local curvature of space–time is a curve that is 5 metres high and 600,000 kilometres across!

The Earth weighs 6 x 10^{21} tonnes (6,000 million million million tonnes, made up of 10^{52} protons and neutrons). This mass bends the local space–time fabric into a very shallow curve that is roughly one-billionth of the curvature of the surface of our planet — but this curve is still steep enough for things to fall towards the centre of the Earth.

The Moon has less mass, and less gravity, than the Earth — so the curvature of space–time is less at the surface of the Moon than at the surface of the Earth.

our planet, they were about one 10 millionth of the diameter of an atomic nucleus — a very small distance indeed! Only very violent events, like exploding supernovas, or black holes colliding into each other, or a black hole eating a star, will give off gravitational waves that are detectable here on Earth.

Gravitational wave detectors

The current generation of gravitational wave detectors need a big lump of something that will 'ring' like a bell — so they use a rod of metal. When a gravity wave comes through, it will shrink and then stretch space — and so the lump will shrink and then stretch. It is hoped that a wave of vibration will sweep up the lump, bounce off one end and then sweep down again.

Joseph Weber came up with the first generation of gravitational wave detectors when he suspended an aluminium bar. If a gravitational wave swept along the bar and then compressed it and expanded it, waves of vibration would ring inside the aluminium bar, bouncing repeatedly off the ends. In 1969, he claimed that he had detected gravitational radiation, but no other experimenters were ever able to duplicate his result.

The next stage was to perfect and fine-tune this concept of detecting vibrations with three gravitational wave detectors widely separated from each other. The reason for having three detectors is to bypass local effects, like a mild earth tremor or a truck rumbling by. If something simultaneously makes all three detectors, scattered over the surface of the planet, start ringing, then that something is probably a gravitational wave. Another reason is that by timing the arrival of the gravitational waves at the three different detectors, we can then work out their origin.

Today there are three gravitational wave detectors on the planet: two in Rome and Louisiana made of aluminium, and one in Western Australia made of niobium.

The gravitational wave detector in Western Australia is a 1.5-tonne lump of niobium, about 3 metres long and 30 centimetres in diameter, which cost about A$250,000 — and it's the largest lump of niobium in the world! It's cooled so close to absolute zero (only 4 Kelvin) that the atoms have virtually stopped vibrating. This means that you can pick up vibrations that are much smaller than the diameter of the nucleus of an atom.

Niobium has a very special quality — it 'rings' better than practically any other material on Earth (apart from silicon and sapphire, but you can't get them in big enough lumps). If you hit a lump of steel with a hammer it will vibrate for a few seconds and then the sound will die away. But this bar of niobium will keep on ringing for weeks. After 17 years of tinkering, scientists have fine-tuned this device to be sensitive enough to detect a pin being dropped in London on the other side of the planet! It's so sensitive that if the Earth, while making its orbit around the Sun, was shifted suddenly by a distance equal to one-thousandth the thickness of a human hair, this niobium bar would start ringing!

The third generation of gravitational wave detectors is called LIGO (Laser Interferometer Gravitational-Wave Observatory). This detector has two stainless-steel pipes, each 4 kilometres long, mounted at right angles to each other. Inside, there is a

EINSTEIN WINS!

Einstein's Theory of General Relativity made a number of predictions, and a few of them have already been found to be true. One of them involves the planet Mercury, the planet closest to the intense gravitational field of the Sun.

The orbit of Mercury is not circular but slightly elliptical (or egg-shaped). The part of this ellipse that is closest to the Sun is called the perihelion. Each time that Mercury orbits the Sun, its perihelion shifts along in its orbit just a little bit from where it was last time.

Our astronomers tell us that the perihelion, or point of closest approach, will itself take 3 million years to make a complete orbit of the Sun. Newton's Law of Gravity does not predict this at all, but Einstein's theory of Gravity does — to an accuracy better than 1 per cent!

Another prediction that Einstein's Theory of Gravity (or General Relativity) made was that light would be bent by a gravitational field.

So if a beam of light from a distant star passes through a strong gravitational field (like that of the Sun), the beam will follow the local curvature of space–time, and bend. In Tahiti, in 1919, during a total eclipse of the Sun, it was found that the apparent positions of various stars close to the line-of-sight of the Sun shifted, just as Einstein's theory predicted. More recent and more accurate observations have confirmed this finding to within 1 per cent accuracy.

Another prediction of Einstein's Theory of Gravity was that an accelerating mass would give off gravity waves. This effect is noticeable only with today's sensitive technology, and only indirectly, and only if the masses are huge. The 1993 Nobel Prize in Physics was won by Joseph Taylor and Russel Hulse, who started work on this topic way back in 1974.

vacuum less than one-hundredth of one-billionth of the Earth's atmospheric pressure. There are mirrors mounted at the end of each pipe, and a laser beam bouncing between the mirrors. If a gravitational wave ripples through it will shorten the distance between the mirrors, and the two laser beams will interfere with each other and show up this difference.

Once it's up and running, this set-up should be 1,000 times more sensitive than the rods of aluminium or niobium.

Neutron stars are (as the name suggests) stars made of neutrons. They are so dense that a thimbleful would weigh some 500 million tonnes. A neutron star weighing 1.4 times as much as our Sun would be only 10 kilometres across (as compared to the 1,500,000 kilometres of our Sun). Some neutron stars are pulsars — they spin very rapidly and squirt out beams of radio waves, a bit like the beam of light from a lighthouse.

Taylor and Hulse decided that an object in the sky called PSR 1913+16 was actually a pair of neutron stars going around each other in a very tight orbit. One of the neutron stars was a pulsar spinning about 160 times a second. The radio beam from the pulsar swept across the Earth, and so gave us a very convenient timing pulse.

The two neutron stars were orbiting around each other every 7.75 hours. Because they were orbiting they were accelerating around a central point, and so each neutron star should, according to Einstein, emit gravitational radiation. But if they were emitting energy in the form of radiation, the combined system of the two neutron stars should slowly lose energy. Because the system was continually losing energy, that meant that the stars would gradually get closer towards each other and that their orbit around each other should get a little quicker. According to Einstein's mathematics, each year the stars should get about a metre closer to each other and each year their 7.75-hour orbit should get shorter by about 75 millionths of a second.

The scientists spent 18 years making painstakingly accurate measurements, and came up with a figure that was within 0.3 per cent of that predicted by Einstein!

So even though we haven't *directly* picked up gravitational radiation with an instrument, we're pretty darn sure that gravitational waves do exist.

There is a real purpose to being able to measure gravitational waves. If we can measure them then we might later be able to control them.

A gravity battery

We can store electrical energy in an electrical battery, but imagine being able to store gravitational energy!

At the moment, the Space Shuttle needs a few thousand tonnes of fuel to launch it into orbit. The energy in that fuel gives the Space Shuttle a whole lot of gravitational potential energy. But this gravitational potential energy is just wasted as heat when the Space Shuttle re-enters the atmosphere.

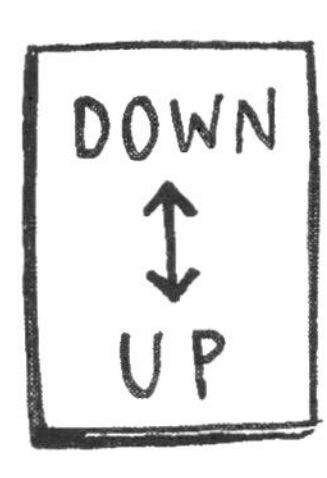

But if, as we gently brought the Space Shuttle down, we could somehow store this enormous amount of energy in a 'gravity battery', we could then use this stored energy to lift the Space Shuttle back into orbit. The only energy we would need would be to overcome frictional losses. Suddenly, space travel would become incredibly cheap.

That's the dream. And though we can't directly measure gravitational waves, if we had something like a gravity battery, the very force that keeps us down might be the pick-up we're looking for.

REFERENCES

Anderson, Ian, 'Australia tunes into gravity waves', *New Scientist*, no. 1987, 22 July 1995, p. 24.

Cline, David B., Mann, Alfred K. & Rubbia, Carlo, 'The search for new families of elementary particles', *Scientific American*, January 1976, pp. 44–54.

DeWitt, Bryce S., 'Quantum gravity', *Scientific American*, December 1983, pp. 104–115.

Faber, Scott, 'Gravity's secret signals', *New Scientist*, no. 1953, 26 November 1994, pp. 40–44.

Freedman, Daniel Z. & van Nieuwenhuizen, Peter, 'Supergravity and the unification of the laws of physics', *Scientific American*, February 1978, pp. 126–142.

Gribbin, John, 'A theory of some gravity', *New Scientist*, no. 1705 (Inside Science no. 31), 24 February 1990, pp. 1–4.

Hecht, Jeff, 'Exotic pulsar blazes trail to glory', *New Scientist*, no. 1896, 23 October 1993, p. 4.

Reader's Digest Book Of Facts, Reader's Digest (Australia), 1994, pp. 396, 398.

EXPLODING STARS & LIQUID MIRRORS

The Milky Way is a big place. It's about 100,000 light years across and has around 400 billion stars in it. But with the naked eye you can see less than one-millionth of 1 per cent of these stars! If you want to see more you need a mirror — the bigger the better. But big mirrors are expensive, unless you make them out of liquid metal!

The space between the stars is not empty — there's dust and gas in there. And thanks to some secret military satellites that were looking for nuclear weapons' explosions, scientists have discovered a star that blew a giant hole in our galaxy.

Geminga

About one-third of a million years ago and a few hundred light years away from our planet, a star exploded. For a few brief moments, that one star was brighter than all the other stars in our galaxy put together. In those few seconds, it blasted out more energy than all the other 400 billion stars in our galaxy would emit during their entire lives and it blew a hole in our galaxy. Teams of astronomers, using various telescopes in orbit and on our planet, have succeeded in tracking down the culprit.

When that star went supernova and exploded, it blasted off its outer layers as cosmic debris, zipping through space at about 60,000 kilometres per second. The space between the stars is not totally empty — there's a very thin cloud of gas and dust. The expanding shock wave from this supernova swept up this gas and dust before it, like a cosmic broom — and in its wake it left behind a volume of space with much of the gas and dust removed. The expanding shock wave actually carved out a bubble in space!

Back at the site of the explosion, what was left of the star began to collapse because the nuclear fires had gone out. Under the influence of the enormous gravitational forces, sub-atomic particles such as electrons and protons were driven into each other to make neutrons. Soon this star was a ball of neutrons — a neutron star — about 20 kilometres across. But even though the nuclear fires had gone out, the star was still hot from gravitational energy and so it continued to give off electromagnetic radiation.

Around 340,000 years ago, at the beginning of the explosion our ancient ancestors would have seen a star suddenly become brighter than the full Moon. That star would have probably been visible in the daytime for two years, before it began to fade.

The light from that explosion crossed the few hundred light years to our solar system in a few hundred years. Much of the

STARS & SUPERSTITIONS

People have always had superstitions about stars.

In the Bible, the Magi (the three wise men) were guided in their search for the new Messiah by the star shining in the East.

In some societies, a meteor (a shooting or falling star) was believed to herald the birth of a new baby. This baby would shortly be born under where the shooting star was seen (not a star, of course, but a much smaller rock).

However, other societies believed meteors were an omen of death.

Either way, falling stars were powerful. So if you were to wish upon a falling star, perhaps some of that power would come to you and make your wish come true.

Most societies believed that counting stars would be unlucky. According to some English superstitions, the act of counting a hundred stars could bring death, and the mere act of pointing at a star could mean instant death.

YOU ARE HERE

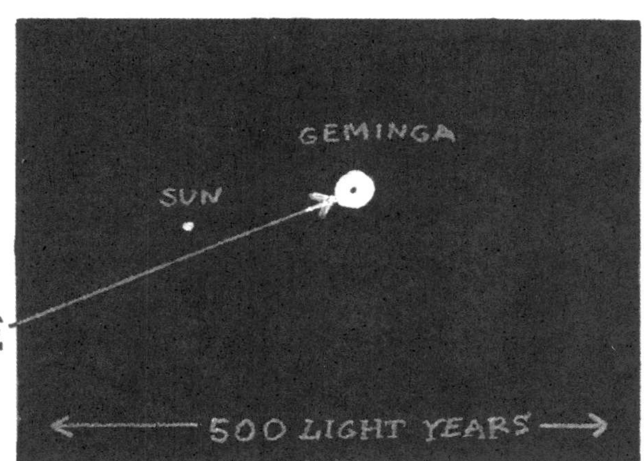

TINY HOLE

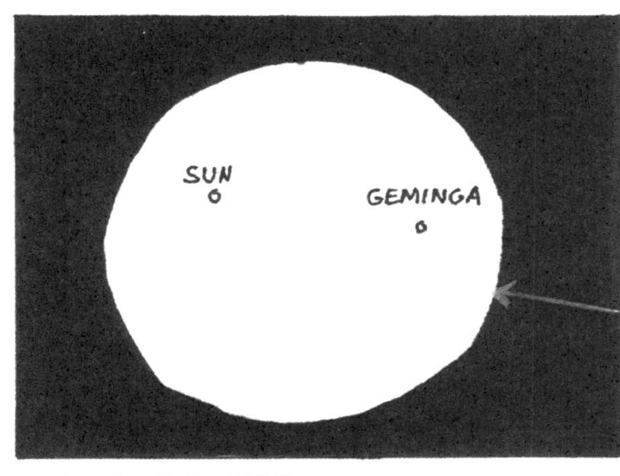

NOW

HUGE HOLE

ABOUT 200 × 600
LIGHT YEARS ACROSS

substance of the star was hurtling outwards from the site of the explosion — but at a much slower rate than the speed of light. As this matter slammed into the gas and dust between the stars, it set up a rapidly expanding shock wave — part of which was heading for our solar system.

But our solar system (the Sun and its nine planets) is surrounded by a bubble of pollution from the Sun. Our Sun vaporises four million tonnes of hydrogen every second. About three million tonnes of this hydrogen are annihilated into pure energy, but one million tonnes appears as charged particles that rush out from the Sun. This one-million-tonnes-per-second makes up most of the pollution bubble that engulfs our solar system.

The original supernova occurred some few hundred light years away. As the shock wave expanded out from the site of the explosion, it slowed down. It took about 9,000 years to cover that distance, and by the time it reached the vicinity of our solar system the shock wave had slowed down to about 1,400 kilometres per second. It smashed into the pollution bubble around our solar system. This pollution bubble normally reaches out from the Sun about three times the distance of Pluto, the outermost planet.

This 9,000-year-old cosmic tsunami of a shock wave pushed this bubble of pollution in towards the Sun — probably as far as Saturn or Jupiter. Then, like a wave pushing around a rock on the seashore, it pushed on by and around our solar system.

It almost certainly left behind some hints of its passage. The shock wave had picked up various chemicals and elements in its travels, and probably dumped them on the airless moons of our outer solar system. But the inner planets like Earth were probably unaffected, because they were protected by the squashed-up pollution bubble protecting our solar system.

The star itself, which had once been brighter than all the other stars in the Milky Way put together, was now a neutron star — an incredibly dense 20-kilometre ball of neutrons. Over the next third of a million years, it gradually became dimmer. Even though it gave out hardly any visible light, it gave out some X-rays and a lot of gamma rays. (Gamma rays are just another form of electromagnetic radiation, but they are tens of thousands to trillions of times much more energetic than visible light.)

Between 1963 and 1970, the Americans launched six pairs of secret military satellites in the Vela series. These Vela satellites were designed to look for nuclear weapons that were violating the Nuclear Test Ban Treaty by being exploded on the surface of our planet. The satellites did this by looking for the characteristic burst of gamma rays that are released when a nuclear weapon explodes. Because they are so energetic, these gamma rays will travel through cloud and rain.

The Vela satellites did not detect any unauthorised nuclear weapons tests on Earth — officially. (But on 22 September 1979, when the satellites were well and truly beyond their seven-year lifespan, Vela 6911 picked up a sudden burst of gamma rays coming from the Indian Ocean. At the time, the Carter administration would not believe the evidence of the ageing Vela satellite. But 18 years later, the South African Deputy Foreign Minister Aziz Pahad admitted that South Africa, with the help of Israel, had made and exploded a nuclear weapon in the Indian Ocean.)

However, during their operating lives, the Vela satellites did find more than a dozen unexplained bursts of gamma rays coming from deep space. These bursts lasted for periods ranging from several thousandths of a second to over hundreds of seconds. Some of this data was eventually declassified in 1973.

In 1972, NASA independently launched SAS–2 — the second Small Astronomical Satellite. During its seven-month life it picked up some 8,000 gamma rays — about one every 40 minutes. It picked up an incredibly bright source of gamma rays in the constellation Gemini. But no matter how hard the scientists looked, they could not find a visible star that corresponded to this object that was giving off gamma rays.

Giovanni Bignami from the Cosmic Physics Institute in Milan called this mysterious object 'Geminga'. He was probably having a joke as well. Officially, he called it Geminga because in *Gemin*i there was a *ga*mma ray source. But in the Milanese dialect, *geminga* means 'it doesn't exist', or 'it isn't there'.

In 1975, a European satellite, CosB, was launched. It also had a gamma ray detector, and it managed to localise the position of Geminga to a patch of sky a bit bigger than the full Moon. Both SAS–2 and CosB had instruments that could pick up gamma rays but were not very precise in saying where the gamma rays came from.

But in 1979, NASA launched the Einstein Observatory (also called the High Energy Aerial Observatory–2) which was designed to pick up X-rays. X-rays are quite close to gamma rays in the energy band. The Einstein Observatory was much more precise than either SAS–2 or CosB in the location of the objects that it tracked. In 1983, Giovanni Bignami announced that he had discovered a source of soft X-rays somewhere close to Geminga.

In 1988, two teams with optical telescopes (which pick up visible light) independently found the star that was Geminga. It was a very blue star, much hotter than our own Sun. Whereas the surface temperature of our Sun is around 5,000 Kelvin, this star had a temperature somewhere between 300,000 and a million Kelvin. Bignami was one of the discoverers, using the New Technology Telescope at the European Southern Observatory in Chile. The other team included Jules Halpern of Columbia University, and Stephen Holt of NASA. They used the 200-inch Palomar Telescope in California to find the same star as Bignami. This star is some 100 million times fainter than anything you can see with the naked eye, and is one of the faintest stars visible to our optical telescopes.

In 1991, the Gamma Ray Observatory, a 16-tonne monster of a satellite, was launched by NASA specifically to look for gamma rays. It soon logged some 2,221 gamma ray photons coming from the direction of Geminga.

In 1992, Halpern and Holt used the newly launched ROSAT X-ray satellite to pick up 7,636 X-ray photons. They analysed them very carefully, and found that Geminga was putting out X-rays in pulses that were 237 milliseconds apart.

Immediately, all of the other scientists who had collected any data in the past relating to Geminga went racing back to their vaults to re-analyse the data on their computer tapes. Now that they had this magical number of 237 milliseconds, they were able to go looking for, and find, a

DISTANCE OF GEMINGA

We think that Geminga, this neutron star remnant of an ancient supernova, is probably about 300 light years away from us. The astronomers think this is the case because Geminga moves so rapidly across the sky, and because the X-rays coming from it are not absorbed by the interstellar gas. If Geminga was further away, we would be getting far fewer X-rays from it.

repeating cycle of roughly this period in the old data. But more importantly, because the data spanned over a decade of time, they were able to measure that Geminga was actually slowing down its rate of pulsing. And so they were able to work out that Geminga was some 340,000 years old.

So now we know that Geminga, the star that blew a hole in space, is the *closest* neutron star — just a few hundred light years away. It puts out lots of gamma rays, some X-rays and hardly any visible light.

We also know that Geminga exploded about a third of a million years ago and created a local bubble in space around us, roughly pear-shaped, about 200 light years by 600 light years. But in the intervening period since it exploded, Geminga has moved so fast that it is practically out of the bubble that it created all that time ago. Geminga glides Northeast across the sky at the rate of 0.17 arc second each year, which means it would cover a distance in the sky roughly equal to the width of the Moon every 10,000 years.

It was lucky for us that Geminga was so far away when it exploded. Otherwise the incredible radiation from the supernova could have wiped out life on our planet, and we wouldn't be here today to talk about it. But perhaps the radiation from Geminga did affect our ancestors — maybe it was bad for the human race of 340,000 years ago, or maybe it was good and stimulated an incredibly rapid evolution in brain size.

BUBBLE IN SPACE

Why is the local bubble, which was blown in the dust of our galaxy by Geminga, pear-shaped and not round?

Our galaxy is in the shape of a flattened spinning disc, like two dinner plates stuck face-to-face. The dust and interstellar gas are most dense in the plane of the galaxy. So the local bubble is narrowest (200 light years) in this plane because of the resistance to the expanding shock wave caused by the density of the gas and dust. But the gas and dust are less dense in a direction up-and-down out of the plain of the galaxy, and it is in this direction that the local bubble is longest (600 light years).

Visible stars

If you're in the Australian Outback at night, you'll notice two things about the night sky — it's very big, and it's full of stars. After you've been staring at the Milky Way and the rest of the sky for a while, you'll start to wonder just how many stars you can actually see. Well, it might seem like millions, but it's closer to a thousand!

If a star is close enough, and bright enough, you will be able to see it with your naked eye. Astronomers measure the brightness of stars in Magnitudes. In a fairly dark suburb on an average night, the faintest star you can see registers at Magnitude 5.5.

So, if you look up a standard astronomical reference like *Sky Catalogue 2000.0*, volume 1, you'll find a table showing how many stars of each Magnitude of brightness you can actually see. According to this modern and highly respected text, there is a total of 2,862 stars visible down to a brightness of Magnitude 5.5. Of course, these stars are surrounding the entire globe of the Earth, so you'll be able to see only about half of these stars above your local horizon at any given time — say, about 1,400 stars.

Some 150 years after the birth of Christ, a Greek astronomer called Ptolemy actually did the experiment at Alexandria, an Egyptian city founded by Alexander the Great. He counted 1,022 stars in the night sky. That's a bit different from 1,400. How come there's a difference?

In 1994, in the magazine *Sky & Telescope*, John Holtz described how he came up with a figure that was closer to Ptolemy's figure. He fed into his computer a database of stars, and then took account of a whole lot of different factors.

For example, there's an effect called 'atmospheric extinction', which means that as a star gets low in the sky and close to the horizon it gets lost from view because the light has to travel through more atmosphere. He was able to program his computer to not count some stars when they got close to the horizon. He was also able to take into account the effect of where the observer is on the surface of the Earth. For example, depending on where you are, some stars are visible all night, other stars can be seen only for a short time after they have risen above the horizon, while others never rise above the local horizon. In Australia, observers can always see the Southern Cross; however, observers on the other side of the Equator can see the Southern Cross only at certain times of the year, and then only for a few hours.

Using his computer program, John Holtz found that, back in the year AD 150, at Alexandria, which is 31 degrees North of the Equator, Ptolemy should have been able to see 1,095 stars, which is very close to the 1,022 that he reported seeing. Then Holtz did the same calculation for 1995, seeing how many stars were visible at different distances from the Equator and how this varied with different times of the night.

According to his calculations, the greatest number of stars is visible an hour or so after sunset. From where he was living, in Pennsylvania, which is about 40 degrees North, he could see his local maximum of around 850 stars just after sunset, dropping to 600 stars visible around midnight. In fact, he found that his latitude of 40 degrees North had the lowest number of stars visible anywhere on the planet!

Holtz also found that Australian observers can see more stars than anyone

MIRROR OR LENS?

Practically all the small amateur telescopes use lenses to bend and concentrate the light, but the larger telescopes all use mirrors. Why is this so?

There are two major problems with using lenses in telescopes.

First, a lens bends the different colours to different degrees — so blue light will be bent more than red light. This will give you a fringe of colours around the object you're looking at.

The second problem is that a big lens (a metre or so) will be so large that in the middle, where it is not supported, it will bend.

It was that great physicist, Isaac Newton, who recognised this problem in the 17th century. He designed and built the reflecting telescope which used mirrors to concentrate the light. Today, this telescope is called a Newtonian telescope.

So you can't have really big lenses, because they sag under their own weight in the middle. But you can have really big *mirrors*, because they can be supported from behind.

else on the planet — about 950 after dusk, dropping to about 800 in the early morning. That's just one of the many special things about the Australian Outback.

Liquid mirrors

Mirrors are pretty amazing devices. Not only can you see your own face in them, but you can put a few mirrors in a tube and make a telescope that tells you how old the Universe is. Unfortunately, big mirrors are expensive. But one astronomer reckons you can make a 2.5-metre mirror for US$20,000 — less than the price of a new, middle-size car — with a slowly spinning bowl of mercury!

People have always wanted mirrors, ever since they saw their own reflections in a pool of still water. Around 5,500 years ago, mirrors of polished bronze were popular in what is now the Middle East, while mirrors of brass are mentioned in the Bible. Some 328 years before the birth of Christ, the Greeks set up a school that taught people how to make different types of mirrors. Some people claim that, by 100 years after the birth of Christ, the Romans had glass mirrors that had a reflective layer of silver behind the glass.

A modern use of mirrors is in telescopes. Astronomers call their telescopes 'light buckets'. The bigger your light bucket, the more light photons it can catch and the deeper you can look into space. An average amateur telescope might have a mirror with a diameter of 15 centimetres. A serious amateur telescope might have a mirror with a diameter of 40 centimetres. But the diameters of professional astronomical mirrors begin around 1 metre, and go up to around 4 or 5 metres. At this stage you are talking tens of millions of dollars just for the mirror.

The shape of a telescope mirror — or for that matter, the reflector in your car's driving lights — is not the shape of a sphere, but of a parabola. A parabola will bring incoming parallel rays of light to a single point — very handy if you want a sharp photo. While it's fairly easy to grind a mirror to a spherical shape, it's complicated and expensive to grind it to a parabolic shape. But by a marvellous coincidence, when you spin a liquid in a container the surface of that liquid takes on the shape of a parabola. The faster you spin, the deeper the parabola.

Ermanno F. Borra, Professor of Physics at Laval University in Quebec, has been making telescopes with spinning liquid mirrors made of mercury for over 10 years. In fact, a half-metre, spinning, liquid-mercury mirror was used to look at the Moon way back in 1908, but the technology of the time made it impossible to have a very smooth surface to the mirror. The footsteps of a horse, even 50 metres away, rippled the face of the spinning liquid mercury!

But Professor Borra had the advantage of modern technology. The bearing he used to hold up the bowl of mercury was lubricated with compressed air instead of oil — so it gave good insulation from vibration coming from the ground. He had the bowl spinning with a very smoothly running electric motor, via a floppy rubber belt. This isolated the mirror from the vibrations of the motor. He then ran into a

MERCURY

Mercury is so dense (13 times denser than water) that iron, rocks and even lead will float on its surface. It's rarer than uranium, but more common than gold or silver. It's also one of the 20 or so 'native elements' (so called because they can be found in a native or uncombined state). It's also the only metal element that exists as a liquid at room temperature.

Mercury has been found in a 3,500-year-old Egyptian tomb. It was used as a red pigment and as a medicine long before the birth of Christ.

Today it's used in mercury switches, mercury batteries, thermometers, in percussion caps for ammunition, and to extract gold. It's also very toxic as a vapour, and it's infamous for causing Minamata Disease. In the town of Minamata, on the Japanese island of Kyushu, mercury poisoned the fish in the bay and then the townspeople who ate the fish.

MAKE YOUR OWN MIRROR

You can use the liquid-mirror technique of Ermanno F. Borra to make your own mirror at home. All you need is some sort of smoothly spinning turntable, a container and an epoxy. For home use, the best epoxy is one that is thin or runny (so that it will easily take up a parabolic shape) and has the longest possible setting time. This mirror, once it hardens, could be covered with reflective Mylar. You couldn't really use this mirror in a telescope, but it'd be perfectly fine for concentrating the heat of the Sun.

problem with the wind currents generated by the spinning mercury — they rippled the mercury.

But he found that a thin film of a plastic called Mylar smoothed out the mercury without distorting the light. And when he checked the surface of his spinning mercury mirror, he was astonished and delighted to find that it had a superb optical surface. It was smooth to about 20 billionths of a metre. That means that if the mirror was as big as the Earth is wide, the biggest bump would be only 1 millimetre high!

However, there's one disadvantage to a liquid-mirror telescope — it always has to point vertically upwards, otherwise the mercury will spill out over the edge. But he found he could take a series of quick electronic snapshots of the sky as his telescope was slowly rotated by the Earth, and then electronically combine them.

The biggest liquid mirror so far is 3 metres across. It's in a telescope to search for orbiting particles of space junk as small as 1.5 centimetres across.

So, thanks to modern technology, it's easy to build a very cheap, and very high-quality, large, liquid-mirror telescope. The old saying goes that the eyes are the mirrors of the soul. With these new ultra-cheap, ultra-high-quality mirrors, you can have your very own 2.5-metre mirror to gaze into the soul of the Universe.

REFERENCES

Borra, Ermanno F., 'Liquid mirrors', *Scientific American*, February 1994, pp. 50–55.

Dragovan, Mark & Alvarez, Don, 'Making a mirror by spinning a liquid', *Scientific American*, February 1994, pp. 86–97.

Fisher, Arthur, 'Spinning scopes', *Popular Science*, October 1987, pp. 76–79, 101–102.

Henbest, Nigel, 'The invisible star shows up', *New Scientist*, no. 1359, 26 May 1983, p. 538.

Hickson, Paul, *et al.*, 'UBC/LAVAL 2.7 meter liquid mirror telescope', *The Astrophyscical Journal*, vol. 436, 1 December 1994, pp. L201–L204.

How Is It Done?, Reader's Digest (Australia), 1990, pp. 128, 180–182, 245, 406.

Iannotta, Ben, 'Spinning images from mercury mirrors', *New Scientist*, no. 1986, 15 July 1995, pp. 38–41.

Why In The World?, Reader's Digest, 1994, p. 219.

AURORAS

You've probably heard of the aurora, and you might have seen pictures of the eerie beauty of giant pink and green curtains, thousands of kilometres long, dancing across the night sky. And if you were in the Australian Outback south of Alice Springs in October 1989, you would have seen an aurora with your naked eyes.

God's TV set

The aurora is God's TV set. In an ordinary TV set, an electron gun squirts electrons at a special screen that glows in various colours. In the upper atmosphere, 100 to 400 kilometres above the ground, charged particles from the Sun ram into atoms of oxygen and nitrogen, which glow in unearthly greenish-white and reddish colours.

You've probably heard of the three states of matter — solid, liquid and gas. You know that water can exist as a solid block of ice, a pool of liquid water, and as a gas (which we call steam). In these well-known states of matter, each atom has a central core that is surrounded by electrons. But 99 per cent of the Universe is not made of solid, liquid or gas — it's made of plasma. Plasma is a very hot state of matter in which the electrons have been ripped from the individual atoms, so the cores of the atoms float in a sea of electrons. You see plasmas in flames, lightning bolts, plasma cutters in your local car-body repair shop, and of course, in the Sun. This plasma is so hot that at the surface of the Sun, about 1 million tonnes of charged particles are blown off into space each second.

The Sun is a giant hydrogen bomb. It throws its fallout, the exhaust gases of this mighty engine, out into the Solar System. This fallout is called the 'solar wind' — a dilute plasma of charged electrons, cores of hydrogen atoms, and other stuff. The solar wind usually blows past the Earth at about 400 kilometres per second (about 1.5 million kilometres per hour). You might have seen proof of the solar wind when Halley's Comet visited in 1986. The tail of Halley's Comet always pointed directly away from the Sun. That's because it was blown away from the Sun by the power of the solar wind.

The Earth is surrounded by an enormous magnetic bubble the shape of a giant teardrop. On the side of the Earth closest to

JUPITER BIGGER THAN MOON!

Our Earth is about 12,800 kilometres across, and the magnetic bubble that surrounds the Earth is about 6 million kilometres across. The magnetic bubble around Jupiter is much bigger again. If we could see the magnetic bubble of Jupiter, it would look bigger than our Moon!

Not only does Jupiter have a much stronger magnetic field than the Earth, it has four major moons, a small moon and a dust ring inside its magnetic bubble. So there's a constant movement of charged particles between the moons and the dust ring. Most of the charged particles are squirted out by volcanoes on the moon Io. Because Jupiter's magnetic bubble is so big and powerful, these charged particles can be accelerated to very high energies. These charged particles are so powerful that they would kill an unprotected human in a few minutes! Any spacecraft that goes close to Jupiter has to be specially hardened against this radiation hazard.

COLOURS OF THE AURORA

The incoming beams of electrons and other charged particles from the Sun smash into atoms in our atmosphere. These different atoms all have electrons revolving around a central core. The electrons are excited by the energy of the colliding particles and jump up to a high energy level, and then, as they drop back down to their normal energy level, give off light.

The most common colour seen in auroras is a whitish-green light (wavelength of 557.7 manometers), which comes from oxygen atoms. Nitrogen atoms give off a beautiful pink colour. There are many other radiations given out, such as ultraviolet and infrared, but we can't see these with the human eye. But we know, from spacecraft that carry cameras that can see in the ultraviolet light, that a typical aurora gives off as much as 10 times more ultraviolet light on the sunlit side than on the dark side.

the Sun, this magnetic bubble is squashed up by the solar wind so it's quite close — about 60,000 kilometres away from the Earth. But on the side away from the Sun, it's stretched out by the solar wind so that it reaches some 6 million kilometres downwind. This bubble protects the Earth from high-speed charged particles from outer space. However, this bubble has major holes in it at the Earth's North and South Magnetic Poles.

There are three main scenarios for auroras.

First, some of the solar wind gradually leaks into the magnetic bubble, which stretches and then suddenly snaps back, injecting charged particles into the upper atmosphere near the North and South Magnetic Poles, creating two simultaneous auroras.

Second, the corona, the Sun's outer layer, can squirt out giant bubbles of plasma, which are entangled with twisted magnetic fields. When these bubbles fly off into space they leave behind dark holes in the corona, which cameras that are sensitive to X-rays can see. These enormous bubbles fly through space at around 800 kilometres per second, and when they hit our upper atmosphere they can set off spectacular auroras.

AURORA FOR THE MATHEMATICIANS!

The power of an aurora is proportional to the product of: (a) the speed of the solar wind, (b) the strength of its magnetic field, and (c) the fourth power of the sine of half the polar angle (as measured from the North Pole) at which this magnetic field hits the Earth's magnetic field.

Finally, the Sun can squirt out a super-giant blob of plasma, at a temperature of about 1 million degrees Celsius and weighing about a billion tonnes, at the amazing speed of 100,000 kilometres per second. A blob this big has enough energy to boil a lake hundreds of times the size of Australia. The Sun squirts out these blobs several times per month, but luckily they mostly miss us.

Magnetic storms

A little ray of sunshine can be a real pain in the aurora.

You might think that space, up where the satellites orbit, is quiet and peaceful — with maybe a hint of a Viennese waltz in the background. But things are actually quite tough for satellites. Activated atoms of oxygen nibble away at their skins, they get

blasted by space junk (which includes urine and faeces from astronauts and cosmonauts), and giant magnetic storms blow up their delicate electronics.

When a powerful gust of the solar wind from the Sun hits the Earth's magnetic field, it makes an enormous generator that turns the energy of the solar wind's motion into electricity. In an average magnetic storm, the electricity generated is roughly equal to the amount of electricity being consumed by the USA at any moment — about a million megawatts. A big storm can be a million times more energetic again!

Magnetic storms interfere with power lines by inducing enormous sparks. In fact, gigantic currents can surge through the very bedrock of our continents and interfere with transoceanic cable telephone lines. Magnetic storms interfere with satellite and radio communications.

The radiation levels in space can be enormous. Your average chest X-ray will give you a radiation dose of a few thousandths of a rad. But in space, satellites can be buffeted by radiation levels as high as 150,000 rads. To protect a satellite's computer chips against these levels of radiation, you'd have to coat the whole satellite with 35 centimetres of aluminium — which would be enormously heavy. So the computer-chip makers build in capacitors to let radiation-induced charges dissipate more rapidly, or they add resistors to slow down certain critical circuits, or they coat certain areas with a layer of a special oxide only 20 billionths of a metre thick. All this adds cost.

In 1997, you could buy a megabyte of RAM (Random Access Memory) for your computer for only A$10, but space-hardened RAM cost $1 million per megabyte. How strange that our most expensive RAM is sent away into space.

These storms can affect humans. According to *New Scientist* magazine,

SINGLE EVENT UPSET

Whenever an industry comes into existence, it develops its own jargon. This jargon, for some unknown reason, usually has lots of TLAs (Three Letter Acronyms). In the space industry, one of these TLAs is a SEU (single event upset).

An SEU happens when a computer circuit has a 'bit flip'. Inside a computer, all information is stored as either a 1 or a 0. When a bit flip happens, a 1 turns into 0, and vice versa. A bit flip will nearly always change a message.

When Voyager 1 had its flyby of Jupiter, it had 42 SEUs. One of them unfortunately reset the clock of the central computer, so the clock was then running a few thousandths of a second behind real time. A few thousandths of a second are no big deal down here on Earth, but they were significant when you consider the very precise movements that Voyager had to go through as it zipped past Jupiter.

But today's spacecraft have been fitted with SEU-resistant computers, which have radiation-hardened chips and shielding.

Russian cosmonaut Sergei Avdeyev was terrified by a magnetic storm while orbiting in the space station Mir. He said: 'I felt that the particles of radiation were walking through my eyes, floating through my brain, and maybe clashing with my nerves.'

In March 1989, the most violent magnetic storm for 30 years hit the Earth. Not only did it bring down the Hydro-Quebec power system in Canada for nine hours, it also knocked out two communications satellites. The cost for this damage alone was US$700 million. This magnetic storm hit our atmosphere with such a huge amount of energy that our atmosphere temporarily expanded further into space, so causing extra drag on low-orbiting satellites. One NASA satellite lost 500 metres of altitude in just one day.

In 1994, another magnetic storm knocked out the Canadian communications satellites, Anik E1 and E2 — on the same day! As a result, essential TV broadcasts, newspaper payouts, credit-card transactions, paging signals and so forth were all lost.

On 6 January 1997, the Sun spewed out an enormous magnetically charged cloud of helium and hydrogen. It rushed towards the Earth at 1.6 million kilometres per hour, and had expanded to some 50 million kilometres across by the time it ran into the magnetic bubble around our planet on 10 January. It dumped some million amps of current into our magnetic bubble. Telstar 401, an AT&T television relay satellite, had a massive electrical discharge and died on 11 January. Some scientists were convinced that this magnetic storm knocked out Telstar 401, but other scientists were more cautious and said that they weren't too sure exactly what caused it to fail.

At the moment, all that we have to warn us are a few research satellites that send us

MAGNETIC STORMS KILL SKYLAB

Magnetic storms do more than just give us incredibly magnificent auroral displays and fry delicate electronics. They can also heat up the atmosphere, which is what killed Skylab.

As the upper atmosphere heats up, it becomes more dense. At 400 kilometres above the ground the density can increase by a factor of 20, and this increased density can increase the drag on a satellite.

Skylab, which was supposed to be NASA's first permanent space station, orbited at around 435 kilometres above ground level. It was used for a few missions, and then abandoned. The intention was that a few astronauts would return to Skylab in the Space Shuttle, and boost it safely to a higher orbit. However, the first launch of the Space Shuttle was delayed for a few years. In July 1979, two years before the Space Shuttle roared into the sky and after the Sun had been particularly violent, Skylab crashed to Earth over Esperance in Western Australia.

back research data a few times each day. In October 1995, the WIND spacecraft happened to record an enormous magnetic storm heading for the Earth at over 3 million kilometres per hour — just before the satellite was due to send a report. That report gave us 30 minutes' warning to shut down major power lines and essential satellites, and saved us millions of dollars. But we don't have working satellites that send us back data full-time.

So now, at the end of the 20th century, our sophisticated technologies are still at the mercy of the Sun. The Sun follows a regular cycle of activity, and the next peak is predicted for around the year 2000. Let's hope that by then we will have learnt how to predict magnetic storms, so Australia can broadcast the Olympics without any glitches. After all, imagine missing out on the last 10 seconds of a gold-medal win!

REFERENCES

Anselmo, Joseph C., 'Solar storm eyed as satellite killer', *Aviation Week & Space Technology*, 27 January 1997, p. 61.

Covault, Craig, 'New explorer to study Aurora', *Aviation Week & Space Technology*, 19 August 1996, p. 31.

'Forecasting the northern lights', *Science*, vol. 268, 28 April 1995, p. 496.

Kerr, Richard A., 'Does the Sun trigger outburts from the Earth's magnetosphere?', *Science*, vol. 271, 15 March 1996, p. 1496.

Muir, Hazel, 'Watch out, here comes the Sun', *New Scientist*, no. 2015, 3 February 1996, pp. 22–26.

Nash, Madeline, 'Cosmic storms coming', *Time*, 9 September 1996, pp. 68–69.

Powell, Corey S., 'The blustery void', *Scientific American*, August 1996, p. 14.

Russell, C.T., 'The magnetosphere', *In The Solar Wind and Earth*, Terra Scientific Publishing Co. (TERRAPUB), Tokyo, 1987, pp. 73–100.

Syun-Ichi Akasofu, 'The dynamic aurora', *Scientific American*, May 1989, pp. 54–63.

Williams, John, 'Built to last', *Astronomy*, December 1990, pp. 36–41.

Reg

Robofish, Albatrosses & Smart Materials

The albatross probably holds the World Record for the Most Efficient Flying Machine. It can stay airborne for thousands of kilometres and burn up hardly any energy at all because it continually adjusts its wings to give the best possible gliding shape. The wings of the albatross are made of natural, biological 'smart materials'. Dolphins change the shape of their skin to reduce drag, so they too have 'smart skin'.

A new branch of engineering will have to be invented to deal with the marriage of smart materials with transportation. Already we have invented robofish, including Charlie the Robofish and Proteus the Robo Penguin. Who knows where smart materials will take us?

Albatross finds food with statistics

The unique gliding ability of the albatross has fascinated mariners for centuries.

There are about 13 or 14 different species of albatross. The biggest one of them all — the Wandering Albatross — is a pretty remarkable bird. Even though a fully grown Wandering Albatross may have an

ALBATROSS RANGE

Some of the first, and even today the best, information about where albatrosses hang out came to us from Captain C.C. Dixon. He was at sea in southern latitudes from 1892 to 1919. He chalked up some 3,500 days of observations on the Great Albatross. This bird spends most of its time between 30 and 60 degrees South, although a few Great Albatrosses have crossed the Equator into the Mediterranean.

enormous wingspan, greater than 3.5 metres across, its body will be only about 25 centimetres across and it will weigh only about 11 kilograms. Not only will it travel up to 4,000 kilometres to get a feed for its baby, but it even uses a strange chaotic form of statistics to help it find that food!

Daddy albatross will usually fly into one of the small islands in the South Pacific or South Atlantic Ocean, close to Antarctica, in late November. Mum will arrive a week or two later.

The egg arrives in December and usually hatches in March. For the first day the newborn albatross chick eats nothing at all, but for the next three days it swallows small amounts of a special oil from the stomach of its parents. Not only is this oil very rich

ALBATROSS, THE FERRARI OF SEABIRDS

Pierre Jauventin and Henri Weimerskirch from the National Centre for Scientific Research at Beavoir have fitted six Wandering Albatrosses with lightweight radio transmitters.

In a single feeding trip, the birds can cover up to 15,000 kilometres. Not only do they have an enormous range, they're speedy. Over the 'short' distance of 800 kilometres, they can average 56 kilometres per hour.

They're patient as well. One albatross sits on the egg to hatch it, and then relies on its foraging mate to bring back food. The one on the nest will stay there for 50 days, before abandoning the egg.

in vitamins A and D, but an albatross can use it as a weapon, squirting it for a metre or more at an attacker.

For the first month of life, the baby albatross cannot control its own body temperature and so the parents take turns sitting on the baby to keep it warm. One parent will fly out to sea, swoop down to swallow some fish or squid, come back home and vomit up this delicious, partly digested mush so the baby can have a feed.

After this first month, both parents go out chasing food for the baby and so it will be home alone. Over the next 250 days, the chick will eat a total of about 80 kilograms of pre-digested fish and squid. By this time the baby is ready to leave the nest for good, which is just as well because Dad is ready to glide in, with stars in his eyes, to mate again.

The albatross is a rather clumsy bird on the ground. Its legs are so far back on its body that it has difficulty in walking. But once it gets into the air, it can stay aloft on the roaring winds of the Sub-Antarctic for hours at a time without flapping its wings even once.

The albatross may be one of the most perfect gliding birds, but it has to land on the water to get food — either for itself or for its baby. And this leads to a problem: what is the best random search pattern to find food, when the food is scattered in small patches over hundreds of thousands of square kilometres of featureless ocean?

One search pattern that you often see in nature is called the Drunkard's Walk. Here, each step can be in any direction but the size of the step is nearly always the same.

But there is another type of random pattern called a Lévy Flight, after the French mining engineer and mathematician Paul-Pierre Lévy. In a Lévy Flight, the steps can vary from really small to really large — quite different from the Drunkard's Walk, where the step is usually the same size. The Lévy Flight is what the Wandering Albatross uses as it sweeps over the oceans around Antarctica.

This was discovered by Peter Prince, a biologist at the British Antarctic Survey in Cambridge, and his colleagues. They wired up five Wandering Albatrosses with electronic

DRUNKARD'S WALK

Drunkard's Walk is the popular name for what the mathematicians call Brownian Motion. In 1827, the Scottish botanist Robert Brown, when looking through a microscope, saw tiny particles moving in an irregular motion. They could go in any direction, but the size of the step was nearly always the same.

In 1900, Louis Bachelier tried to use the mathematics behind Brownian motion to study changes in the prices of the stock market. In 1905, Albert Einstein came up with a mathematical model for the movement of particles that Robert Brown first saw. Einstein's work helped show that the then-unproven molecular theory of matter was correct. In 1926, Jean Perrin, who did experiments following Einstein's work, won the Nobel Prize for physics.

JAPANESE ALBATROSS

In 1798, Samuel Taylor Coleridge's famous narrative poem *The Rime of the Ancient Mariner* was published. The story is told by the ancient mariner, who commits a very foolish crime when he kills a friendly albatross and the ship's good luck turns bad. As part of his penance, he has to wear the albatross around his neck. So today the word 'albatross' can mean 'an obstacle to one's success'.

In Japan, the golfers call a hole-in-one an *arubatorosu*, an albatross. To celebrate their success in getting a hole-in-one, they have to buy food, drinks and presents for any other members of the golf club who happen to be around, and for their friends as well! This can easily cost over US$13,000.

So in Japan, some 4 million golfers spend about US$280 million a year in hole-in-one insurance — to guard *against* the possibility that they'll get a hole-in-one.

activity recorders and radio transmitters, and watched them do 19 foraging trips. The albatrosses would fly a great distance, swoop down to catch some food, have a few more short hops to catch more food, and then make another long journey.

If you do the mathematics, it turns out that a Lévy Flight search pattern can cover much more ocean than a Drunkard's Walk and it's also much better at finding the scattered patches of squid and fish that the albatross needs.

So even though they're called the Wandering Albatross, they're actually using a very efficient search pattern.

Robofish

Fish have been swimming in the seas of our planet for hundreds of millions of years, so it's no surprise that they're really efficient at it. Engineers know that there's no point in re-inventing the wheel, so they've started copying the supremely efficient designs of nature. That's how come we have a 'tuna' called Charlie — the first robofish — swimming in a tank at the Massachusetts Institute of Technology.

This research is being carried out by Michael S. Triantafyllou, Director of the Ocean Engineering Testing Tank Facility at

ALBATROSS DEATHS

On Sand Island in the Midway Atoll, some 30,000 Laysan Albatrosses were killed by the US Navy during World War II in an effort to reduce collisions between military aircraft and the birds. However, the navy stopped killing the albatrosses once it was realised that the albatrosses loved certain sand dunes which gave them favourable updraughts. Once these special sand dunes were levelled, there were 70 per cent fewer collisions!

MIT, and his brother George, who is Professor of Mechanical Engineering at the City College of New York. There are many good reasons why they're trying to copy fish.

Your average agile fish can do a U-turn without slowing down, whereas a ship has to slow down by over 50 per cent to reverse its direction. And fish can do their complete U-turn in a very tight curve, with a radius only about one-fifth the length of their bodies, while ships need to have a curve 10 times larger. A very manoeuvrable ship could avoid many accidents.

Dolphins are remarkable ocean-going creatures. They can generate about 4 horsepower and can travel at 80 kilometres per hour — but if we want a boat to travel at that speed, we have to give it a 70 horsepower engine! Each year, billions of tonnes of goods are moved across the oceans by large ships — if we could increase their efficiency, it would mean an enormous saving in fuel bills.

And of course, there are the military applications. Propellers are noisy, and the Navy would like quiet ships.

These reasons — manoeuvrability, efficiency and low noise — are just a few of the reasons why Charlie the Robofish was invented. He has a complicated mechanical skeleton, with over 2,000 different parts. He's about 1.25 metres long, has some 40 ribs, a segmented flexible spine connected by six aluminium hinges, and tendons made of thin stainless-steel cables that run over pulleys and are driven by six electric motors. His skin is made of a layer of foam and Lycra which is smooth enough not to have any wrinkles or bulges (which would cause extra turbulence).

A real fish has a sense organ called the 'lateral line' arranged along its sides. This organ can measure the pressure in the water along the sides of the fish's body as it travels along. Being a reasonable copy, Charlie the Robofish has pressure sensors there as well. And, just like a real fish, he's pushed through the water by a tail that flaps back and forth, just like the shape and action of a real tuna tail.

Why is a fish much more efficient at moving through the water than a ship with a propeller? Well, a propeller generates a long spinning vortex or whirlpool that stretches behind the ship, robbing energy and slowing the ship down.

But the flapping tail of a fish is quite different. As it sweeps to one side, it generates (say) a clockwise rotating vortex in the water. But when the tail flips back in the other direction, it generates an anticlockwise vortex. The next step is the key to the efficiency of the fish.

The anticlockwise vortex runs into the clockwise vortex and when they meet they squirt water away from the fish. Newton said, 'For every reaction there is an equal and opposite reaction', and so, if the water squirts backwards, there's a reaction that pushes the fish forwards.

It turns out that, in all fish that have been looked at so far, there is a careful match of the following: (a) the frequency of tail flapping, (b) the side-to-side distance covered by the flapping tail, and (c) the speed at which the fish travels. These are carefully 'chosen' by the fish so that the clockwise vortex always runs into the anticlockwise vortex.

Because of this marvellously simple mechanism, fish use hardly any energy to move through the water. Even an early artificial fish like Charlie the Robofish can achieve 86 per cent efficiency, while the best

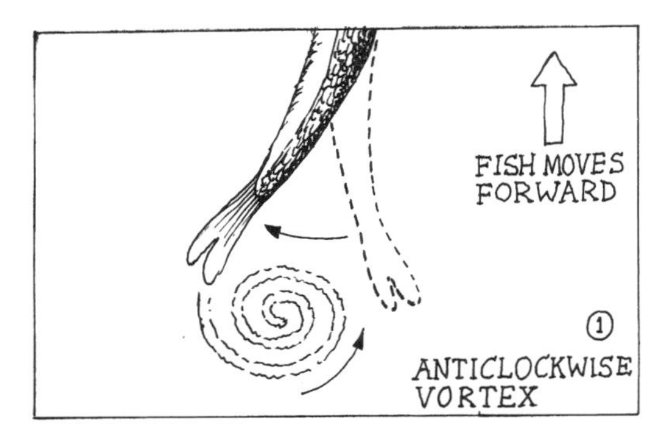

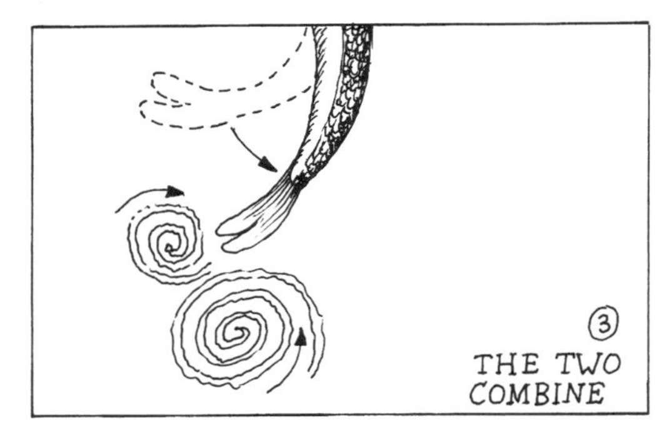

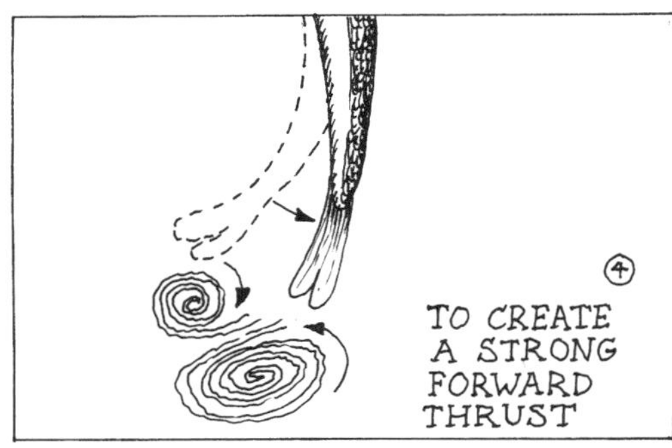

that our ships can achieve is 40 per cent. Of course, fish use extra tricks besides just flapping their tails. They also flex their bodies, so they can even extract energy out of the vortices that roll down the side of their bodies. At the moment, making an exact copy of a fish, with a smooth and continuously bending body, is way beyond what we can build into the robots of the late 1990s.

But now Charlie the Robofish has been superceded by Proteus the Robo Penguin.

Robo penguin

Proteus the Penguin has been built by Professor Triantafyllou and James Czarnowski, a graduate student. One day, while James was watching penguins at the New England Aquarium, he suddenly realised that penguins, with their fairly plumpish bodies, have a shape a lot closer to large ships than the tuna has.

So Proteus the Penguin has two penguin-like flippers at the tail end. Each flipper (or foil) looks like a rudder and has its own motor. The motor can give each foil two different motions — one a side-to-side flapping, and the other a kind of twisting motion. The movements of the foils are synchronised by a small personal computer, and the power supply comes from two car batteries. Proteus is 4 metres long and half a metre wide.

As one of these rudder-like foils flaps through the water, it shoves a greater area of water than the blades of a propeller do. So more water is pushed backwards, and there's more thrust forwards. A foil has another advantage over a propeller. The water that has been forced to rotate by the propeller continues rotating for many metres behind the ship. This is wasted energy. But the foils don't make the water rotate, so that there's a smaller energy loss. In the tests so far, the foils have an efficiency of 87 per cent, which is better than the 70 per cent efficiency for a propeller. (The efficiency of the whole ship is less than the efficiency of the propeller.)

Proteus the Penguin has a top speed of 2 metres per second, which it gets by flapping its foils 200 times each minute. But if Proteus was scaled up to the size of a container ship, the foils would flap once every two seconds, giving a total speed of 30 knots. According to the US statistics for 1992, if the efficiency of merchant ships in the American Fleet were improved by only 10 per cent, this would save some US$500 million per year. It would also save some 4 billion litres of fuel each year.

The military would like Proteus the Penguin as well. Propellers sound like propellers, and all ocean-going powers have records of the 'signatures' of the submarines of other nations. But a propulsion system built on penguin power

would be more likely to blend into the background noise of the ocean.

Once we learn a few more fish secrets, we can put the 'fish' back into 'ef*fic*iency'!

Smart materials that self-repair

We humans are really good at making things. But one major problem with manufactured things is that they immediately begin to wear out — and that's because they can't repair themselves. For example, when a car comes out of a car factory, well, that's as good as it's ever going to get.

But things are different with the human body. Consider bone. Weight for weight, bone is stronger than steel. But bone also warns you (via the mechanism of pain) when it's under threat of infection or getting close to its breaking point. Bone can even repair itself.

If only our manufactured goods could sense damage, and repair themselves! It sounds an impossible dream, but already our engineers are inventing a new generation of 'smart materials', and soon we'll have self-repairing materials that have 'feelings'.

You'll understand why this is important, if you think back to 1988, when metal fatigue ripped off the roof of a Boeing 737 in Hawaii. This particular short-haul plane had been flying for 19 years. It had made many thousands of take-offs and landings, on short hops from island to island in the Hawaii group. On this particular flight, when the plane had reached an altitude of 7 kilometres, 6 metres of the roof of the cabin simply broke off and vanished. A flight attendant was sucked out of the plane and plummeted to her death, and 61 of the passengers were injured. When Boeing checked other 737s around the world, they found tiny fatigue cracks in about half of the planes they inspected.

Now, it's hard to find cracks. The first method that technicians use to find cracks in aircraft is by simply having a look. They don't do this after each flight. At regular intervals in the life of plane, they actually have to strip out the entire internal fittings so they can check critical structures. The second method is by blasting ultrasound waves into the metal of the plane, and analysing the echoes that come back. But the sensors that generate the ultrasound are so small (roughly the size of a 50 cent coin), that it would take months to check every single square centimetre on a 747.

Worldwide, airlines spend about $US30 billion each year on the maintenance, monitoring and repair of planes. So wouldn't it be wonderful to make planes out of materials that are sensitive enough to know that cracks are developing, and resourceful enough to repair those cracks?

In the human body, we have nerves that tell our brain that we have bumped into a wall or touched something too hot. Ray Measures and his team at the Institute of Aerospace Studies in Toronto are making non-biological alternatives — they're making materials with optical nerves. They install optical fibres into the structure of a plane as it's being built. They run light through the optical fibres. As the plane's structure is stretched or squashed, the frequency of the light running through the optical fibre changes. This means technicians can continually monitor the state of health of the structures in the plane via this built-in optical fibre.

One advantage of a built-in optical fibre is that the engineers could program the computers to take measurements at millions of points around the plane, thousands of times every second. Unfortunately, even today's fastest computers may not be able to deal with this huge flood of information.

But we could use a trick that the human body uses. You don't feel your shirt touching against your chest, or your socks against your feet, because your body decides that there's no point in monitoring these sensations all day. In the same way, we could teach the computers to ignore these signals if they were always the same, and to signal an alert only when something happens.

Of course, we would have to make sure that not only would the optical sensors be reliable and consistent in recording data, but also that they wouldn't weaken the structure in which they were installed.

The new smart materials in planes can do more than just feel that damage is about to happen — they can actually repair the damage as it happens.

Carolyn Dry, at the University of Illinois at Urbana-Champaign, is inventing materials that have special liquid chemicals stored inside hollow fibres. When the material develops a crack, this crack also breaks open the hollow fibres, and the liquids leak out and repair the damage.

Cliff Friend at Cranfield University in the United Kingdom is following a different approach. He uses alloys which can actually change shape. When the material develops a crack, these built-in alloys will change their shape and close the crack.

The orthopaedic surgeons, or bone doctors, are fond of saying that 'bone is the prince of tissues'. And they're right — bone is the only tissue in the body that will repair itself to an as-new condition. Maybe in the near future, with this new technology, both bones and Boeings will be able to repair themselves!

Smart materials that change shape

There are a million uses for materials that will change shape.

If you've ever sat in a passenger jet just behind the wings, you'll have noticed the flaps coming out of the back of the wing to make the area of the wing bigger as the jet comes down to land. But this is crude technology.

Consider a plane such as a Boeing 747. Its wings have been shaped to give the best possible fuel economy when the plane is travelling at cruising speed. But suppose that there is a delay at the airport and the plane has been put into a stack, constantly circling at a low speed. Under these conditions, the wings are much less fuel-efficient and can use up to a quarter times more fuel. Imagine that the wings could change to a shape that would be more efficient at a lower speed!

Ron Barrett, from Auburn University in Alabama, headed a team which designed, built and flew a small aeroplane which they called Mothra. Mothra was the very first aeroplane with smart structures built into the wings. It didn't have any of the conventional aerodynamic control surfaces flaps and ailerons. Instead the tail plane and the tail wings actually change shape. This is done via piezoelectric actuators. ('Piezo' means rock and 'electric' means electric.

SMART MATERIALS TO CONTROL VIBRATION

An aeroplane flying at high speed generates a lot of noise because the panels on the fuselage vibrate. If we could make these panels so smart that they could, under their own local control, vibrate with a movement that exactly compensated for the externally forced vibration, the panel would be completely still and no noise would get into the plane's cabin. Already, we have made one primitive type of self-vibrating panel — skis have been developed which will change their shape slightly for different conditions, to allow the skier to go faster.

Helicopters could benefit from this technology to control noise and vibration. Helicopters have all sorts of dynamic forces acting on the blades of their rotor. The faster the helicopter flies, the bigger are these vibrations. Above 430 kilometres per hour, the vibrations are so great that you begin to damage the helicopter. If you go any faster, you begin to damage the helicopter. Smart materials could counteract these vibrations, and let the helicopter fly faster.

You find piezoelectric crystals in spark guns for lighting your gas oven. When you squeeze the rock, it gives off electricity. It can also work backwards. When you put in electricity, the rock can either squeeze or expand.)

In their Mothra plane, Barrett and his team fed electricity into small piezoelectric actuators, which expanded or shrank, depending on how electrical current was put into them. Unfortunately, these piezoelectric actuators simply don't have enough grunt to control a full-size plane.

There is another technology called 'musclewire', which will shrink or expand depending on the sort of electricity you put into it. They do generate enough force to lift a whole kilogram, but they respond only slowly. However, other smart materials can respond very quickly.

At the moment, this entire field of smart materials is a new, young and expanding field. As we get more skilled with smart materials, and also know more about the human DNA, we'll get closer to marrying biology and technology. One day, we might even be able to grow muscles that are fed on glucose, to change the shape of an aeroplane wing.

REFERENCES

'Albatrosses break speed and distance records', *New Scientist*, no. 1707, 10 March 1990, p. 16.

Baker, Howard, 'Pushing the boat out with penguin power', *New Scientist*, no. 2080, 3 May 1997, p. 25.

Friend, Cliff, 'Even aircraft have feelings', *New Scientist*, no. 2015, 3 February 1996, pp. 32–35.

Gray, Sir James, 'How fishes swim', *Scientific American*, August 1957, pp. 48–54.

Stix, Gary, 'Robo tuna', *Scientific American*, January 1994, p. 127.

Tickell, W.L.N., 'The Great Albatrosses', *Scientific American*, November 1970, pp. 84–93.

Triantafyllou, Michael S. & Triantafyllou, George S., 'An efficient swimming machine', *Scientific American*, March 1995, pp. 40–48.

Vincent, Julian, 'Tricks of nature', *New Scientist*, no. 2043, 17 August 1996, pp. 38–40.

MUSIC & GOSSIP MAKE MEGABRAINS

All societies have some sort of music, even though they may not have words for it. But why do we bother having music — after all, it may sound nice, but what use is it? Back in the 1690s the English dramatist and poet William Congreve knew of one use for music. In his play *The Mourning Bride* he says: 'Music has charms to sooth a savage breast, to soften rocks, or bend a knotted oak.' Music might be able to do all that and, according to some research, music might even make you brainier! (And gossip might have already made our brains as big as they are now!)

Did cavemen dig the blues?

One set of musical instruments, discovered in an archaeological dig in Kiev, are 20,000 years old. These six instruments were made from bones from various parts of a mammoth — the shoulder blades, hip, jaw, tusks and skull. According to the musicologists from the Hermitage Museum in St Petersburg, these primitive percussion instruments still held their tone after 20,000 years.

But the current record of oldest musical instrument is held by a 24,000-year-old flute from France. It was made from the bones of a vulture. One strange thing about this ancient flute, as well as many other ancient flutes, is that it was tuned to play 'blue notes'.

What are 'blue notes'?

Practically all of today's music is based on the well-tempered scale (or equal temperament scale), in which there are 12 equally spaced semitone intervals in each octave, and each note is 5.9 per cent higher

MUSIC & ANIMALS

Animals are supposed to sometimes have a special relationship with music. Some farmers think that playing music to cows while the animals are being milked helps the milkflow. In the movie *Bringing Up Baby* a leopard loves the song 'I Can't Give You Anything But Love, Baby'.

Back in 1989, Gadi Gvaryahu, from the Hebrew University in Jersualem, found that playing quiet calming music to chickens and turkeys helped them gain about 3 per cent extra weight. The music also seemed to make them happier.

Jaak Panksepp, a psychobiologist from Bowling Green State University in Ohio, has spent over a decade looking at how animals and people respond to various noises and music. He found that he could get chills to run up the spines of men and women by letting them hear recordings of their children crying. He also found that chickens had a similar response to Pink Floyd. When he played some chickens 'The Final Cut', they ruffled their feathers, which is the chicken equivalent of a shiver running down the spine.

In ancient times, the Roman writer Aelian claimed that you could charm stingrays out of the water by playing the pipes and dancing. He also claimed that you could attract stags and boars by playing 'the sweetest melodies possible'.

Even John Wesley, who founded the Methodist religion, did an experiment on the effects of music on tigers and lions. He went to the lions' den at the Tower of London on New Year's Eve in 1764, and took a flute player along with him. He later wrote in his journal that one of the lions 'came to the front of his den and seemed to be all attention'.

in frequency than the note immediately below it. When jazz musicians (such as Roland Kirk and Sidney Bechet) flatten the third and seventh notes (that is, when they make the intervals smaller), these notes — which have a melancholy sound — are called blue notes. So I guess you *can* accurately say that the cavemen dug the blues!

More recent music

All societies seem to have some kind of music, even if they don't have a name for it. To play music, you need an instrument.

A 5,000-year-old bone flute, made from the bone of a deer, was found in a cave approximately 50 kilometres North of Lisbon in Portugal. This cave was a necropolis, where dead people were laid down in specific patterns. There were some 150 skeletons in the cave, and accompanying them were objects such as axes, adzes, awls, bracelets, shells and musical instruments. So even back then, music was valued.

It seems that Western music began in the so-called 'fertile crescent' at the eastern end of the Mediterranean. The peoples who lived there some 5,500 years ago had all four types of musical instruments, according to pictures and instruments that survive. Probably our oldest surviving piece of music, many thousands of years old, is the Shaduf Chant. This chant is still sung by irrigation workers on the Nile River, as they use their muscles to lift buckets of water from the river onto the fields.

According to the 1993 *Guinness Book of Records*, the world's earliest surviving musical notation dates back to around 1800 BC, or about 3,800 years ago.

The musical tradition made its way from Greece to Rome and then into the rest of Europe. The Christian religion was closely linked to the development of music in the West. The early Christians would chant prayers and scriptures, usually in Latin. Their songs often didn't have a regular rhythm, but would instead follow the accents of the pronunciation of the original

MUSIC & MADNESS

There has long been a belief that music can calm an angry or mad person. In 1913, the great filmmaker D.W. Griffith made a short one-reel movie called *The House Of Darkness*. In this movie, a madman has violent urges, which can be calmed only by piano music.

MUSIC & SPEECH

Music has to be different from the other ways of expressing yourself. Maurice Ravel was the famous musician who in 1928 composed the *Boléro*. He became ill in his later years. We don't know exactly what diseases he suffered from, but they destroyed his ability to speak and write.

But for a full year after he could neither speak nor write, he was still able to compose music. So the centres in the brain for music were, in him, more robust than the centres for speech and writing.

TYPES OF INSTRUMENTS

In the past, there were four basic types of instruments: membranophones, aerophones, idiophones and chordophones.

Membranophones (e.g., drums) make sounds from skins that are stretched over a resonant cavity. Aerophones (e.g., bugles, harmonicas) make sounds when air is blown into a cavity (which is often long and cylindrical) and resonates. Idiophones (e.g., rattles, xylophones) make sound when all, or most, of the body of the instrument resonates. Chordophones (e.g., pianos, harpsichords, guitars) have strings, which make sounds when they are struck or plucked.

With our modern technology, we have a new type of instrument, the electrophone (e.g., synthesizer, electronic organ), which uses electronics to make sounds.

Latin words. Pope Gregory I compiled and organised much of these songs, so they came to be called 'Gregorian Chants'. For a long time, Western music was basically religious songs.

Musicians invented time!

Back in the Middle Ages, music helped us understand this strange thing that we call 'time'. Before then, time was thought of as something that related to the outside world, such as the rising of the Sun.

But back in the Middle Ages, it was the musicians who first realised that time was a separate entity, unrelated to the physical world.

Before then, time was thought of as something that related to the Sun. For example, you might have arranged to meet somebody a bit before sunset, or perhaps at midday, or so on. Today we have clocks that are actually more accurate than the spinning of the Earth, and the situation is now turned around — so we can say that sunset happens (for example) at 6.42.03 pm. But back in the Middle Ages, it was the musicians, with their new-found skills of musical composition, who realised that time was a separate entity.

In fact, an unknown person (who was probably a student at the University of Paris, and who has been given the name of Anonymous IV) wrote a book about music, in which he had a whole chapter devoted to 'rests'. He called them 'a pause or omission of sound for a definite length of time or durational units'. This new musical emphasis on pure music (music without words),

EXPENSIVE STRINGS

Musical instruments can fetch astonishing prices. In 1990, a Fender Stratocaster that originally belonged to Jimi Hendrix was sold by his former drummer, Mitch Mitchell, for £198,000. In that same year, a 1720 'Mendelssohn' Stradivarius was sold for £902,000. Both of these stringed instruments were sold to anonymous buyers.

The highest price ever paid for a piano was $US390,000. This Steinway Grand was built around 1880, and was sold to a person who could not play the piano.

rather than on religious music (music and words praising God), caused a lot of opposition from the Church. In fact, Pope John XXII, early in the 14th century, wrote a Papal Bull in which he strongly condemned those who were 'greatly concerned with the measurement of time values'.

But because these musicians were so concerned with understanding time, a scientific revolution happened. Three centuries later, Isaac Newton could write: 'Absolute, true and mathematical time, of itself, and from its own nature, flows equably without relation to anything external.' Thanks to the musicians, time was seen to be something that was not necessarily related to the physical world, but happened by itself — and that great revolution in thought came *not* from the scientist, but from the musicians!

Learn music and grow smarter

I began thinking about the power of music when I was at a prize-giving night for the Science Faculty at the University of Sydney. As student after student came up to collect prizes, and as his or her accomplishments were described, I was amazed at how the vast majority of the prize-winning science students were also musicians.

It seems that music might have an odd side-effect — music builds up your brain!

One three-year study in Switzerland looked at 1,200 children, and found that children actually learn faster when they have more music classes. These children, aged between seven and 15, were split among some 70 different classes. Their classes ran for 45 minutes.

Before the study began, all 1,200 kids had one or two music lessons a week. But once the study had started, half the kids had the number of music lessons increased to five every week. And to balance up this increased time doing music, these kids had fewer classes in maths and language.

The results, after three years, were astonishing. The kids who were given the extra music classes were actually better at language, and as good at maths, as the other kids — even though they had had fewer language and maths classes than the other kids. The music-class kids were also better in retelling stories that had been read to them — whether they wrote the stories in words or drew them in pictures. And the kids that had the extra music classes also were quicker at learning how to read.

So what is going on? We all know that exercise of your body will make your body grow stronger. Well maybe learning to play

MUSICAL MALADIES

Listening to music might be good for your brain, but playing music might be bad for your body.

Hunter Fry, an Australian physiologist, interviewed the players in seven orchestras. He found that 75 per cent of the string players had various 'industrial' diseases. There seem to be four main groups of problems that affect musicians: 'over use' syndromes, cramps, psychological stress and squashed nerves.

Each different type of musician had their own preferred illness. The names of their diseases included flautist's chin, cello scrotum, fiddler's neck and guitar nipple.

Dr Eric LePage, the Senior Research Scientist at the National Acoustic Laboratories in Sydney, said that most classical musicians in orchestras had a hearing loss. The hearing loss was least with wind and string players, greater with brass players, and huge with percussion players. He said: 'in just over a three-year period, we can see a significant change in the amount of ear damage of percussion players.'

music is the right exercise to make your brain grow smarter. Even learning something trivial, such as how to play 'Chopsticks' on the piano, could force your mind to stretch itself in such a way that the whole brain benefits.

Listen to music and improve your spatial IQ

Another very interesting result came out in 1993, from Frances H. Rauscher and her colleagues from the Centre for the Neurobiology of Learning and Memory at the University of California at Irvine.

As you know, your IQ (Intelligence Quotient) is a rough measure of how smart you are. There are different types of IQ: verbal IQ (how good you are with words), mathematical IQ (how good you at numbers), spatial IQ (how good you are with shapes), and so on.

Rauscher dealt with spatial IQ. Her scientific paper claimed that if students were given the task of organising cut-out, folded paper shapes into various patterns, they did a lot better when they listened to Mozart's Sonata for Two Pianos in D Major, K488. This effect lasted only for about 10 minutes, but while they were listening to the music their performance was better.

Other types of music, such as technopop or compositions by Philip Glass, did not improve the students' performance at arranging shapes. Being better at arranging bits of paper is not the same as being smarter, but it is an interesting result. Looking at shapes and patterns, and rearranging them, is important in a whole range of activities. These activities include playing chess, engineering, drawing and building.

Violin-playing makes your brain bigger

Yet another study showed that playing the violin might make your brain grow bigger.

When you're playing the violin, your right arm (the bow arm) provides intricate balance and precision as it moves the bow back and forth over the strings. But the fingers of your left hand have an even harder task. They have to press exactly on the right spot, and on the right strings, with the right pressure, and for exactly the right amount of time — and to do all this very accurately and often quite rapidly. This study found that the part of the brain that controls the fingers of the left hand (which is usually the weaker hand) is actually bigger in violin students, as compared to everybody else.

Of course, it could just be that people who have a bigger part of their brain to control their left fingers will somehow gravitate to violin-playing, but that doesn't seem likely.

Music and surgery

Dr Lawrence Golden has told the American Psychosomatic Society of his study, which showed that pleasant music can help patients survive surgery better.

His study had 40 patients, who were having out-patient surgery for either glaucoma or cataracts. Half the patients had the choice of listening to music of their own choice through headphones during and after the surgery, while the other half did not listen to any music at all. The patients who were able to choose their own background headphone music were less likely to have their heart rate and blood pressure increase during the operation, and later reported that they were less anxious during the procedure.

FRACTAL MUSIC

In the movie *Amadeus*, the Emperor Joseph of Austria seems to enjoy hearing the world premiere of Mozart's *Abduction from the Seraglio*. However, the Emperor complains that the music has 'too many notes in it'! Well, fractally compressed music may have cheered him up.

Kenneth Hsü, a geologist from the Federal Technical University in Zurich, and his son Andrew Hsü, a musician from the Conservatorium and Music University of Zurich, have invented a fractal-compression technique for music. It doesn't work for all music, but it works for Mozart and Bach. When they used fractal-compression on J.S. Bach's Invention No. 1 in C Major, the music was still perfectly recognisable when they removed half or even three-quarters of the notes. The altered version of Bach sounded 'like Bach but with an economy of trills and ornamentation', according to the Hsüs.

And when they shrank it down to one 64th of the original, they were left with just three notes — which are the fundamental three notes of the entire composition.

MOVE TO THE GROOVE

How come when you hear loud music that you enjoy, you want to get up and dance?

A psychologist from the University of Manchester, Neil Todd, thinks that it's something to do with the other function of the human ear. We all know that the human ear is for hearing. But its other job is keep our balance. One part of the balance-keeping system is called the sacculus. Neil Todd was surprised to find that the sacculus, which is mainly involved with keeping your balance, is also sensitive to sound. It responds to sounds of above 70 decibels, so it will definitely respond to music levels above 100 decibels.

At the moment, we don't really know what happens to a sacculus when you stimulate it with loud music. But it may well be part of the reason that music makes you want to dance.

In this case, the music is better than drugs — it's very cheap, it works, and it doesn't have any bad side-effects.

Music sculpts your brain

Thinkers and composers have thought deeply about music. In his book *Reflections on the Human Condition*, the American philosopher Eric Hoffer says: 'It is the stretched soul that makes music, and souls are stretched by the pull of opposites — opposite bents, tastes, yearnings, loyalties. Where there is no polarity — where energies flow smoothly in one direction — there will be much doing but no music.'

Frank Zappa, the famous rock musician, once said: 'Music, in performance, is a type of sculpture. The air in the performance is sculpted into something.' Well, maybe music also sculpts your brain.

Gossip is good

The front pages of our newspapers and magazines are full of gossip and rumour, whether about politics or the private lives

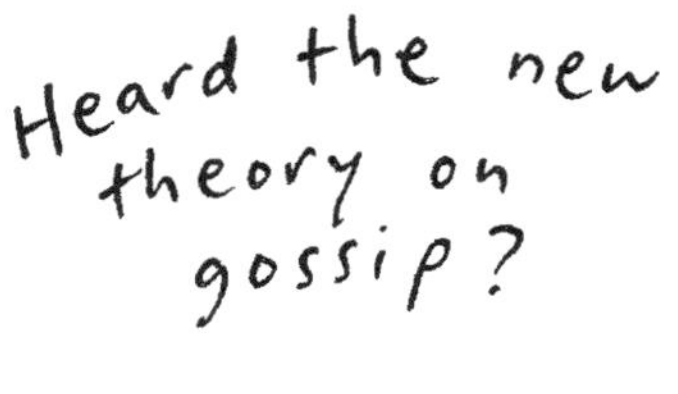

of the rich and fatuous. There must be something very important about this gossip, because humans keep on buying the newspapers.

Robin Dunbar, a Professor of Biological Anthropology in London (the home of Royal Gossip), has come up with a theory to explain why we love to listen to scandal. His unprovable, but interesting, theory links together language, brain-surface size, and the 'natural' size of a community. His theory deals only with primates — two-legged mammals, such as lemurs, chimpanzees and humans.

We humans, compared to other primates, have a very large surface area on our brains. This surface layer is called the neocortex, and it deals with those things that make us uniquely human — poetry, Scud missiles and gossip. Dunbar reckons that the size of an animal's neocortex is related to the size of its community.

Madagascan lemurs, who have a small neocortex, hang around in groups of 10. Chimpanzees, who have a bigger neocortex, hang around in groups of about 50. From those and other numbers, he reckons that humans, with the biggest neocortex, should hang around in communities of around 150 — and history seems to prove this.

Villages in the Middle East about 8,000 years ago had populations of 120 to 150. Even today, the Hutterites of North America live in communities of 60 to 150 people. These religious fundamentalists live on collective farms and reckon that, provided a group is under 150 people, the group can be brought into line just by peer pressure. So whenever a group gets over 150 people, it splits into two smaller groups. (Incidentally, my phone book lists 160 people whom I regularly call.)

The second part of Dunbar's theory says that, as communities got bigger, language simply had to evolve. All the primates spend a lot of time interacting with each other. Lemurs and chimpanzees sit around in a circle and pick off each other's ticks and lice. This takes up 20 per cent of the day for chimps.

In a bigger group of 150 humans, this social grooming would take up 50 per cent of the time — there'd be no time left over for anything else.

BLACK & WHITE HEARING LOSS

Black people are slightly more protected from hearing loss arising from loud music, compared to white people, according to Professor Steve Jones of the University College in London.

He says that this is nothing to do with racist or supremacist theories. The only significant difference between black and white people is the amount of melanin in their skin. He says that experiments show that after both black and white people have been listening to noises of the same level, black people will recover their hearing more quickly. He says that perhaps they suffer less hearing damage because the melanin in the ear soaks up the energy generated by the sound.

According to Professor Dunbar's theory, as the group got bigger, language was invented as a time-efficient way of social grooming and interaction. With hands alone, you can pick only one chimp's lice, but with language you can interact with several others at the same time. When you're picking lice, all you can do is pick lice, but with language, you can talk and walk and gather berries, all at the same time.

And with language, you can build up ties with people who are not around you by talking *about* them — in other words, by gossiping about them. According to Professor Dunbar, gossip was very important in tying the groups of early humans together.

This new theory is quite different from the old theory, which claims that language was invented by males as they went out hunting. In fact, it's far more likely that language was invented by females. Even today, all the psychology tests show that women have better language skills than men.

So in human evolution we've moved from nit-picking to gossip, but with controversial theories like this one we've probably moved back to nit-picking again.

REFERENCES

'Black and white of loud music', *Sunday Telegraph* (Sydney), 16 June 1996, p. 38.

Chapman, Mardi, 'Musicians risk hearing loss', *Australian Doctor Weekly*, 31 May 1996, p. 16.

de Carvalho, Rui, 'Did Stone Age people dance to the flute?', *New Scientist*, no. 1959, 7 January 1995, p. 9.

Dunbar, Robin, 'Why gossip is good for you', *New Scientist*, no. 1848, 21 November 1992, pp. 28–31.

Edwards, Rob, 'Children learn faster to the sound of music', *New Scientist*, no. 2030, 18 May 1996, p. 6.

Elbert, Thomas *et al.*, 'Increased cortical representation of the fingers of the left hand in string players', *Science*, vol. 270, 13 October 1995, pp. 305–307.

Guinness Book of Records, CD-ROM, Guinness Publishing Ltd, 1993.

Hamer, Mick, 'Haunting tunes from ghostly players', *New Scientist*, no. 2048, 21 September 1996, p. 12.

Henahan, John, 'Music soothes patient nerves', *Australian Doctor Weekly*, 11 April 1997, p. 34.

Hoffer, Eric, *Reflections on the Human Condition*, 1973.

Lewin, Roger, 'The fractal structure of music', *New Scientist*, no. 1767, 4 May 1991, p. 15.

Motluk, Alison, 'Just gotta dance, gotta dance', *New Scientist*, no. 2047, 14 September 1996, p. 10.

Rauscher, Frances H. *et al.*, 'Music and spatial task performance', *Nature*, 14 October 1993, p. 611.

Szamosi, Geza, 'The origin of time', *The Sciences*, September–October 1986, pp. 32–39.

Zappa, Frank, *The Real Frank Zappa Book*, written with Peter Occhiogrosso, Picador, 1989.

FOOD POISONING & SQUEEZING

Back in the 1950s, the food that you got on your dinner plate was probably totally sterile and free of germs — meat that had been heated up until it changed colour, and three well-boiled vegetables, each of a different colour. But today our tastes have changed. We don't like our food salted or tinned (processes that were designed to keep bacteria away), and we're moving to lightly cooked, preservative-free lean meats accompanied by barely cooked vegetables. In Australia and New Zealand alone, over 600 people die every year from illnesses related to food poisoning — and between 460,000 and 2.3 million food-related illnesses don't get reported.

But if you squeeze food to a pressure five times greater than the pressure at the bottom of the Pacific Ocean, you can kill the bacteria that cause food poisoning.

Food poisoning

Every year food poisoning kills millions of people around the world. About 3 million children die each year from diarrhoea; of these, 2 million die from food poisoning from contaminated food. Bacteria attack us all the time — bacteria like *Salmonella* in eggs, *Listeria* in cheese and *Escherichia coli* in meats.

Different infective agents such as viruses, bacteria, fungi and parasites can be involved in food poisoning.

But let's just look at bacterial food poisoning. A *New Scientist* article in 1994 discussed a scenario involving the bacteria *E. coli*, *Campylobacter* and *Listeria*. Let's imagine that a food animal drinks water from a river contaminated with these bacteria. These bacteria make their way into the food animal, and once it's killed, into some hamburger mincemeat.

Let's suppose that the food handler in a fast-food shop is keen for a morning break, and he or she irresponsibly puts the hamburger meat next to some soft cheese. Some of the *Listeria* bacteria migrate across to the cheese. The *E. coli* stays in the meat, but a fly lands on the meat, picks up some *Campylobacter* and then lands on the open mouth of a nearby milk carton. After the break, the apprentice cooks the hamburger mince only lightly, leaving the *E. coli* alive and well in the red uncooked centre of the meat.

There are three bacteria involved, so let's discuss three customers.

Let's say the first customer gets the *E. coli*. There are many types of *E. coli*, and most of them are harmless to us. But there are two rather nasty strains of *E. coli*, and these are 0157:H7 and 0111:H (which, in salami, caused an outbreak of food poisoning in South Australia in 1995). A

DRAGON POISONS 65

In January 1996, the Denver Zoo in the USA held an exhibit of Komodo Dragons. This exhibit set off the largest ever reported outbreak of reptile-associated salmonella food poisoning.

The food-poisoning illness lasted for an average of nine days. Of the 65 victims, 38 had bloody diarrhoea and six were admitted to hospital. Twenty-three of the 65 had to be treated with antibiotics.

The four dragons didn't have any direct contact with the public, because of their nasty disposition and because they're considered to be man-eating. The salmonella bacteria was transmitted via the *fence*. The salmonella made its way from the gut of one of the dragons and onto the ground, and then onto the dragons' paws. The dragons rested their paws on the fence from time to time. When the dragons were not nearby, the people would touch the fence. Sooner or later, their hands would touch their mouths and the food poisoning would begin.

customer eats some of this contaminated and undercooked hamburger meat. The *E. coli* makes toxins, which attack his kidneys, and he dies some three days later.

A second customer in the fast-food joint orders a milkshake, which is made from the milk that was contaminated with the *Campylobacter* (courtesy of the fly). Three days later, she begins to suffer massive diarrhoea, vomiting and fever, but she recovers.

A third customer settles for a dish involving the soft succulent cheese which was contaminated with the *Listeria*. He dies of meningitis some three weeks later.

Now, the odd thing is that we all need a licence to drive a car, but there's no licence in elementary food safety for people who handle food. These cases of food poisoning could have been avoided with four simple rules.

First, cook foods properly. Bringing meats to at least 75 degrees Celsius will kill *E. coli* and most of the other bacteria which cause food poisoning.

Second, keep bacteria that live on raw foods, especially the bacteria that live on raw meat, away from foods that have already been cooked, or which, like salads, won't be cooked. This means that once raw meat has been on a chopping board or cut with a knife, you then have to thoroughly wash your hands, the chopping board and the knife.

The third rule is to keep food-poisoning bacteria that live on our body away from prepared foods. For example, 50 per cent of us have a bacterium called *Staphylococcus aureus* living on our skin. A classic case of this type of food poisoning happened on a jet flying from Anchorage to Copenhagen. A food handler on this flight had a pustule on his hand, and this contaminated the passengers' food. Some 200 of the 343 passengers came down with

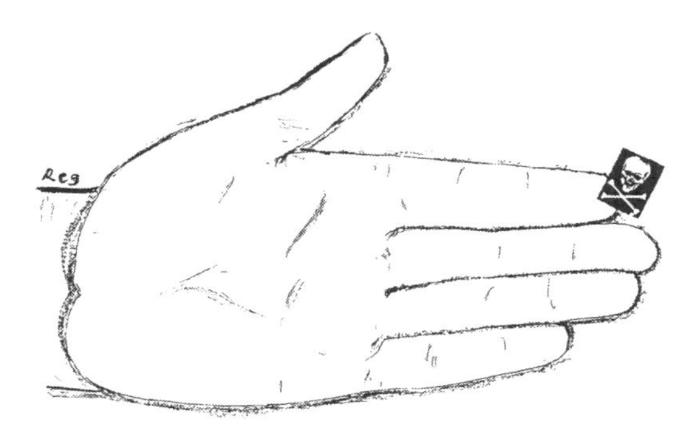

food poisoning, and 30 of those needed treatment with intravenous fluids. So you need good personal hygiene, which includes washing your hands. In fact, you should try to minimise any direct hand contact with prepared foods and use clean utensils instead.

SALMONELLA KILLS GRANNY

On 9 December 1996, Mary McMamee was recovering well from successful minor surgery, which had removed a small ulcer from near her ankle. Even though she was 91 years old, she was in excellent health.

She fasted all day before the surgery, and ate a sandwich after it. She died about a week later from salmonella poisoning, thanks to the sandwich filling.

BACTERIA MUTATE

Antibiotics can kill some bacteria, but over a period of time the bacteria mutate to become resistant to those antibiotics. Because it takes only a day or so before the next generation of bacteria comes along, one generation can very rapidly spread its drug resistance to its descendants.

How do bacteria get resistant to the drugs?

Scientists have found that a high percentage of each generation of bacteria are 'mutators'. They actually have a genetic flaw that makes them mutate — not only can they not repair errors in their own DNA, they will easily take on DNA from other bacteria. The scientists had thought that there would tend to be, in any population of bacteria, perhaps 0.01 per cent of the population as mutators. But they have now found that mutators compose up to 5 or 6 per cent of the population in some strains of *E. coli* and *Salmonella enterica*.

Finally, you should keep prepared foods at the right temperature. Anywhere between 5 degrees Celsius and 60 degrees Celsius will let food-poisoning bacteria multiply rapidly. Once you have cooked food, you must either keep it piping hot over 60 degrees Celsius, or cold under 5 degrees Celsius. If you do reheat cooked food, make sure you heat it to over 75 Celsius degrees. And if you're not sure of the temperatures, food thermometers can be bought cheaply in any shopping centre.

Food poisoning is very common and it's on the increase, but it is easy to avoid once you know what to do.

PARTY ICE CAN POISON

If you get food poisoning after you've been at a party, you'll almost certainly blame the food poisoning on what you *ate*. But a study by the New South Wales Health Department found that some of the pre-packaged *ice* that is sold for use at parties has dangerously high levels of bacteria and heavy metals.

Squeezing food

What can we do about *Salmonella* in eggs, *Listeria* in cheese and *E. coli* in meats? Well, you've probably heard of pasteurisation, so let me tell you about a brand-new technique to kill bugs in food, called pressurisation.

Yes, it sounds unbelievable, but food technologists have found that if they squeeze food up to pressures of 9,000 atmospheres, they can snuff out the bacteria that cause food poisoning.

One of the standard ways to kill bacteria in food is by way of heat. For example, a tin of canned peas will be heated at 120 degrees Celsius for an hour. Sure, this kills the bacteria, but it also wrecks the texture of the peas, and their colour, so that you have to add green dye to bring back their normal colour. The high temperature also breaks down the molecules that give you the flavour, turning them into other molecules that don't taste so good.

But high pressure can kill bacteria without losing the flavour. You don't just put an apple in a vice and squeeze it — you put the food in a pressure vessel and then crank up the pressure.

There's a cell membrane around the cells of every living creature (whether it's a bacterium, broccoli or bilby). The job of the cell membrane is to keep the outside world out, and the inside in. The pressure damages this membrane and makes it leaky. This really upsets bacteria, which are always active, but it doesn't have much effect on the cells in the food, which are basically asleep and just hanging around waiting to be eaten.

The high pressure also damages the cell's DNA and ribosomes, both of which are involved in making proteins. *E. coli* is already infamous for killing a few people in Australia, but a pressure of only 700 atmospheres is enough to stop the DNA and the ribosomes from making any more proteins. Once again, this bothers the bacteria, but not the cells in food.

BUGS, BOWELS & JOINTS

Although food poisoning can be bad enough at the time that you're suffering from it, once it's over it can still affect you!

Even six months after the initial bout, one in four patients can have altered bowel habits. One in 14 can develop Irritable Bowel Syndrome. More rarely, some people can develop active arthritis and may require joint transplants.

In Japan, yomogimochi is a delicacy eaten around the New Year. It's a steamed rice paste that is mixed with a herb. Yomogimochi is normally prepared freshly because the shop-bought stuff just doesn't have the taste of the home-made yomogimochi. Rikimauru Hayashi of Kyoto University has used the high-pressure technique to treat yomogimochi, and finds

SALTY FOOD CAN BE DANGEROUS

The reason that food has been salted for thousands of years is because high salt levels kill bacteria.

Steven Dealler and Richard Lacey, microbiologists from the University of Leeds, have found that microwaves do not easily penetrate salty food. The outside of salty food gets hot, but the inside can stay cool. Any bacteria that have mutated to be resistant to the salt can easily survive the low temperature. According to Dealler and Lacey, the microwaves induce currents to flow in the outer layers of the salted food. These currents then counteract the microwave energy before it can penetrate deep into the food.

FOOD POISONING AROUND THE WORLD

In the USA alone, there are about 10,000 'needless deaths' from food poisoning each year. This is on top of some 50 million non-fatal cases of food poisoning. The annual bill for food poisoning in the USA is somewhere around US$5 billion.

But food poisoning is worse in the Third World. Fritz Käferstein, chief of the Food Safety Unit for the World Health Organisation, said: 'People just don't think of contaminated food as the main cause of diarrhoea and as one of the world's big killers; it's contaminated water and poor sanitation they think of first.'

According to the 1990 United Nations Environment Program, about 45 per cent of the rivers on our planet are already contaminated with dangerously high levels of faecal bacteria such as *E. coli*. Many of the poorer countries use 'night soil' (human faeces) to fertilise their crops, and this can lead to viruses or bacteria entering the food chain. Melons, which suck up huge amounts of water, are notorious for carrying bacteria.

In the wealthier countries, there are slightly different causes of food poisoning. There are many large, centralised food-processing factories. This can lead to cheaper food because of economies of scale, but it can also mean that a single contaminated animal can spoil an entire batch. There has also been a move away from home cooking — in the USA today, about 60 per cent of all food is eaten away from the home.

that it tastes just as nice as the home-made variety. In 1996, the Japanese bought US$1 million worth of this high-pressure delicacy.

The Japanese are the first to adopt this new technique of food preservation. Already you can buy delicious fruit jams and yoghurts, as well as fruit juices. In fact, pressurisation removes some of the bitterness from grapefruit juice, making it more popular. Another advantage of pressurisation is that it leaves vitamin C untouched, whereas it is normally destroyed by the heat process.

It's still early days with this new pressurisation technique. Some foods that have internal air spaces, such as cucumbers, end up as mush if you hit them with high pressure. Grapes become harder, while cabbage gets softer. Mushrooms and potatoes actually go brown because the pressure speeds up the work of an oxidising enzyme.

But this high-pressure speed-up effect can be put to good use. This high pressure also speeds up ripening enzymes, such as rennet in cheese, so cheeses can get to your table sooner — which means cheaper.

And high pressure is good for meat as well. It breaks the connections between proteins such as actin and myosin, which leaves the meat more tender. And the high

FOOD POISONING & FORGETTING

There are thousands of different species of microscopic algae, but only a few dozen of them have toxins that bother us. Shellfish will eat the algae, and the toxins may build up in the body of the mussel, oyster or clam. Shellfish food poisoning can give you diarrhoea, paralysis, and sometimes even loss of memory!

Amnesic Shellfish Poisoning was first described in 1987. The patients brought to a Canadian hospital were all suffering from the classic symptoms of food poisoning, as well a bunch of rather odd symptoms. One of these odd symptoms was a loss of short-term memory, which included the memory of what they had just recently eaten! In this epidemic involving mussels from Prince Edward Island, three people died and over 100 were severely poisoned.

pressure leaves the colour of the meat that lovely delicate pink.

At the moment, this pressurisation technique is popular only in Japan. But already, the US Department of Defence is pouring lots of money into this pressurisation technique to develop field rations for their troops.

So in the future, with regard to food poisoning, you can avoid the anxiety because the food will already be stressed out — just let the food do the worrying for you!

REFERENCES

Collee, Gerald, 'Food Poisoning', *New Scientist*, No. 1687 (Inside Science), 21 October 1989, pp. 1–4.

Dunlevy, Sue, 'Drinkers told to cool urge for party ice', *Daily Telegraph Mirror* (Sydney), 20 February 1997, p. 3.

Grady, Denise, 'Quick-change pathogens gain an evolutionary edge', *Science*, vol. 274, 15 November 1996, p. 1081.

Hill, Stephen, 'Squeezing the death out of food', *New Scientist*, no. 2077, 12 April 1997, pp. 28–32.

James, Tony, 'Poison on your plate', *Australian Doctor Weekly*, 16 May 1997, pp. 28–33.

Johnson, Howard M., Russell, Jeffry K. & Pontzer, Carol H., 'Superantigens in human disease', *Scientific American*, April 1992, pp. 42–47, 73.

Kingsland, James, 'Salty food stays cool to its core in microwaves', *New Scientist*, No. 1713, 21 April 1990, p. 12.

Maurice, John, 'The rise and rise of food poisoning', *New Scientist*, no. 1956, 17 December 1994, pp. 28–33.

Murray, Terry, 'Dragon exhibit linked to outbreak', *Australian Doctor Weekly*, 15 November 1996, p. 47.

'Out of bugs and bowels', *Medical Journal of Australia*, vol. 166, 21 April 1997, p. 442.

'Salmonella kills granny after surgery', *Daily Telegraph Mirror* (Sydney), 19 December 1996, p. 21.

C?
Si.

HEPATITIS C — A SERIOUS THREAT

'Hep' means 'liver', while 'itis' means 'inflammation of' — so hepatitis is a vague word meaning inflammation of the liver. There are many causes of hepatitis — drugs (such as alcohol, methyldopa, paracetamol, isoniazid), toxins (such as from the wild mushroom *Amanita phalloides*) and viruses.

One recently discovered virus that causes hepatitis has already infected 10 per cent of the world's population, and will cost society more than AIDS ever will — and yet the money set apart to deal with this virus is less than 2 per cent of the money allocated to AIDS.

A range of symptoms

The symptoms of hepatitis can range from just feeling a bit off-colour for a little while to having a vague pain in the right upper section of your tummy, through to having a yellow skin and feeling like vomiting, up to getting cirrhosis and sometimes going all the way to getting cancer of the liver.

Alphabet soup

We've known for a long time that people get different types of hepatitis. In the past, it was only by the natural history of your particular hepatitis — how you got it, how sick you got, how long it lasted, and so on — that doctors could work out which type you had.

Around the early 1970s, medical scientists invented tests to pick the virus causing your hepatitis. So 'infectious hepatitis' was relabelled Hepatitis A, while 'serum hepatitis' became Hepatitis B. But our knowledge is increasing so fast, we'll probably end up using half the alphabet!

The virus that causes Hepatitis A is quite small, about 27 nanometres in diameter (a nanometre is one-millionth of a millimetre). You get this virus via your mouth — from food or any other contaminated object. Outbreaks of Hepatitis A happen in army camps and in childcare centres. It's usually a fairly mild disease, but it can sometimes make you very sick. However, it does not usually cause a cancer.

The Hepatitis B virus is a bit bigger, about 42 nanometres in diameter. You mostly get Hepatitis B by direct blood-to-blood contact, such as through sharing needles, or sometimes through sexual intercourse. Hepatitis B is a fairly nasty hepatitis to get, and 1 per cent of all cases progress to cancer of the liver, leading to death.

So back in the 1970s, if you got sick with hepatitis your blood would be tested and in most cases you would be diagnosed as having either Hepatitis A or Hepatitis B. In some cases the patient would definitely have a hepatitis, but their blood was negative for Hepatitis A and Hepatitis B. So this hepatitis, which was neither Hepatitis A nor Hepatitis B, was called Hepatitis 'non-A, non-B'.

It turned out that buried in that convenient label of Hepatitis 'non-A, non-B' was a small alphabet of viruses, all of which cause hepatitis.

In 1977, Mario Rizetto, an Italian doctor, discovered the next hepatitis virus. It's called the 'Delta Hepatitis Virus', or Hepatitis D. It can't replicate on its own and needs the Hepatitis B virus to help it. The Hepatitis D virus has a diameter of 36 nanometres and is a circular, single-stranded, negative-polarity RNA virus. It has caused epidemics worldwide, but it's most common around the Mediterranean, Northern Africa and the Middle East.

There are two main ways of getting infected with a hepatitis virus — either via the mouth (like Hepatitis A), or not via the mouth (like Hepatitis B).

HEP B ANTI-CANCER VACCINE

The vaccine against Hepatitis B is one of the first anti-cancer vaccines on the market. One per cent of all people who get Hepatitis B will get a cancer of the liver that will kill them — so, taking the vaccine will stop them from getting Hepatitis B and the cancer.

In 1984, as a medical student, I took the first generation of Hepatitis B vaccine, which, we were told, was made from the blood of gay men in New York with AIDS. We all figured that the risk of getting something from the vaccine was much less than the risk of getting Hepatitis B in the hospital, so we all took the vaccine.

Soon after, genetic engineering techniques were used to get the vaccine made by a germ living in a stainless-steel vat — so the current vaccines are definitely safe.

I feel that it is the right thing to recommend that all Australian children are immunised against Hepatitis B.

HEP G - ACCIDENTAL TOURIST

Hepatitis G seems to be a very mild virus. It can be transmitted by blood transfusion, but it seems to cause only a mild short-term illness and doesn't seem to have any long-term effects.

Hepatitis G is most often found in people who have other hepatitis viruses, and it's rarely found by itself. So Hepatitis G seems to be an accidental, and fairly harmless, tourist.

But until we know more about it, we should be careful of it.

Hepatitis E, like Hepatitis A, can infect you via the mouth. It's a fairly small virus, about 31 nanometres in diameter. It has caused hepatitis epidemics in India, Africa, Asia and Central America.

Like Hepatitis A and E, Hepatitis F is transmitted via the mouth. The scientists claim to have found small particles of virus, about 27–37 nanometres in diameter, but their claim has not yet been confirmed by other scientists.

Next on the scene were two more hepatitis viruses: GB Virus A and GB Virus B. They were named after a Chicago surgeon known only as GB. He suffered from a hepatitis about 30 years ago, and his blood was stored all that time. These viruses were discovered in April 1996.

In January 1995, Genelabs Technologies announced that they had discovered yet another hepatitis virus, which they have called Hepatitis G.

And finally we come to Hepatitis C. It's a reasonably big virus — almost 80 nanometres across — and it's a major worry. After Hepatitis B, it's the main cause of hepatitis that is not transmitted via the mouth.

As of the late 1990s, we know that there are many cases of severe hepatitis that are not caused by hepatitis viruses A, B, C, D, E, F or G — so we'll very probably work our way further up the hepatitis alphabet.

Hepatitis C

Hepatitis C is a vastly underrated and poorly understood disease. Worldwide, it infects about 500 million people, or about 10 per cent of the Earth's population. In Australia, by mid 1997, 250,000 people had Hepatitis C, and about 6,000 people were becoming newly infected each year.

Hepatitis C was discovered in 1988. The first of the blood tests to diagnose Hepatitis C came out in January 1990, and by February 1990 all the blood donated to Australian Blood Banks was being tested. Before then, blood transfusions and blood products carried some risk, and today some 20 per cent of all the Australians who have Hepatitis C caught it via this pathway.

Most Australians who have Hepatitis C got it by some kind of blood contact with infected blood. The vast majority got it

HEP C & CHINESE HERBS

Some Chinese herbal medicines may actually slow the course of Hepatitis C and reduce the amount of inflammation of the liver. Associate Professor Robert Bailey, of the Gastroenterology Unit at the John Hunter Hospital in Newcastle, carried out a study with 43 patients who had Hepatitis C. Nineteen of the patients were given a placebo (which had no real effect), while 24 were given a Chinese herbal preparation.

One of the markers of liver damage is a chemical called ALT — the higher the level, the more the liver damage. Over a six-month period, the patients on the Chinese herbs had their ALT levels drop significantly, while the patients on the placebo had no change. The only side-effects reported with the Chinese herbs were flatulence and diarrhoea.

It's an interesting study, but the numbers of patients are very small. It would be nice to repeat the study with a much larger group of patients.

by sharing an infected needle — in other words, by injecting drugs such as heroin, morphine and steroids. A few people got it by unsterile tattooing or body-piercing. Others got it by sharing a straw to inhale cocaine. Some people got Hepatitis C via needle-stick injuries while working in the medical field.

It's very rare to pick up Hepatitis C by living with people or by having sex with them. When it does happen, it's usually via blood-to-blood contact, or more rarely, by sharing a razor blade or even a toothbrush.

Hepatitis C affects different people in different ways. Let's look at 100 people who catch Hepatitis C.

About 15–20 will get rid of the virus in the first month, will never be affected by it again and will have normal liver function tests.

About 60 people will have a long-term, or chronic, infection. Some of these people will not be affected by it, and will have hardly any symptoms. Some people will have mild liver damage with mild symptoms. Some will have severe liver damage with severe symptoms, including tiredness, nausea and abdominal pain.

Another 20–25 people will not be so lucky. They will have a long-term infection that will cause, after about 20 years, serious liver damage. Of these people, about 10–15 will remain stable. But the remaining 10 people will be very unfortunate. They will, after another 5–10 years, progress to liver failure or liver cancer.

However, most people with Hepatitis C will have a normal life span.

Each year in Australia, there are about 1.4 million tests done for Hepatitis C.

Nothing made by humans is perfect — and that includes blood tests. So some people with Hepatitis C are wrongly diagnosed to be free of the disease, while other people who don't have the disease are wrongly diagnosed to have the disease.

You can diagnose Hepatitis C with a variety of tests. So far, there have been four generations of the standard Enzyme ImmunoAssay test. Each generation is just a little more sensitive and accurate than the generation before it.

If you are infected with Hepatitis C, you can try avoiding alcohol and anything else that makes your liver work hard. Some people get relief with a drug called ribavirin, while others are helped with a drug called interferon alpha–2b. There are at least six different versions of Hepatitis C that infect people, and some of the versions are more susceptible than others to these drugs.

In Australia, by mid 1997, Hepatitis C had infected 250,000 people — that is, one in every 72 people — with around 6,000 new recruits each year. For HIV infection, the Australian figures at the end of 1996 were about 20,300 people infected — that is, one in every 890 people — and 850 new infections each year. Even though Hepatitis C is less nasty and less virulent a disease than AIDS, it has infected more people — and so the ultimate cost to the community will be greater. Australia has done a magnificent job in AIDS education and awareness. But the amount of money allocated to Hepatitis C is only A$1.9 million per year, as compared to A$110 million for AIDS.

The time has come to think seriously about Hepatitis C.

REFERENCES

'An accidental tourist', *Medical Journal of Australia*, Vol. 166, 21 April 1997, p. 442.

Cristofani, Kathryn, 'Chinese herbs help treat hep C', *Australian Doctor Weekly*, 20 September 1996, p. 1.

'Hep C therapy "disappointing"', *Australian Doctor Weekly*, 18 April 1997, p. 32.

Lamont, Leonie, 'Call for action to halt epidemic of Hepatitis C', *Sydney Morning Herald*, 27 August 1997, p. 4.

Quayle, Sue, 'Hep B immunisation furore', *Australian Doctor Weekly*, 25 April 1997, pp. 54, 56.

Sarzin, Anne, 'Cocaine straws spread hep C', *Australian Doctor Weekly*, 27 June 1997, p. 21.

KILLER PHONES & OTHER MEDICAL MADNESS

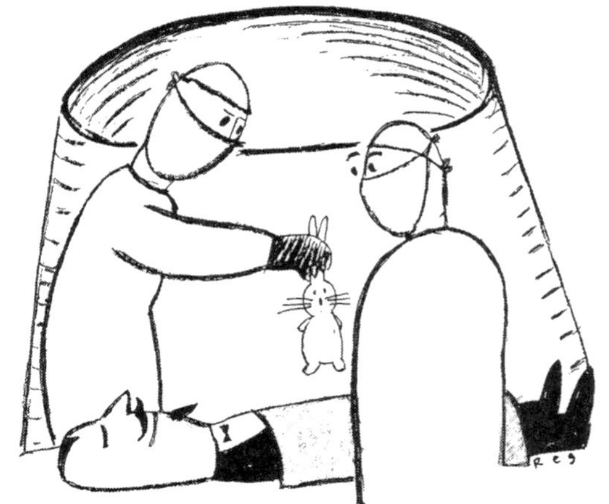

Medicine has a long and honourable history. Healers have great responsibilities, and also great privileges. In the natural course of their work they get to find out all kinds of private information.

Killer phones

In our society, technology is changing very rapidly — and as it does, new diseases change right along with it. Something as innocent as the phone is now being involved in a few diseases.

Consider the case of a 36-year-old woman who spoke on the phone for 32 minutes while she was ironing. By lifting her right shoulder and tipping her head down rather sharply to the right side, she was able to hold the handset between her shoulder and her right ear. At the end of the phone call she was suddenly attacked by a sharp pain on the right side of her neck, which was almost immediately followed by a loud ringing in her right ear.

She was diagnosed at the hospital as having a completely blocked right carotid artery — one of the main arteries carrying blood to the brain. Now, an artery is a bit

like a water hose — it has a fairly thick wall surrounding a hollow interior called the 'lumen', which carries the blood. Somehow a tear developed between the lumen of her carotid artery and the wall. Blood rushed from the lumen of the artery into the wall of the artery, and then expanded the wall of the artery so much that it completely closed off the lumen and stopped the blood flow. To finish things off, the blood turned into a clot.

The woman was immediately given drugs to dissolve the clot, and had to continue taking them for the next three months. The pain in her neck vanished after about 20 hours, but the ringing in her ear lasted for 48 hours. After a few months, the blood clot blocking the lumen of her artery had disappeared and she now has normal blood flow to her brain.

One advantage of a cordless phone is that you can have a long conversation with someone while simultaneously doing other necessary activities around the house. But research shows that a major cause of strokes in young people is exactly this kind of blockage in the internal carotid artery. Maybe we should all wear receptionists' headsets for long phone calls?

Mobile phones can be involved in another disease — the trauma of a car accident.

Driving a car is job that requires your full-time concentration, but many people would like to use this time for making and taking phone calls. By 1995 in North America, cellular phones were so popular that the number of new subscribers to cellular phones was greater than the birth rate! However, a few places, such as Brazil, Switzerland and Israel, as well as some states in Australia, have laws about using cellular phones while you're driving a vehicle.

In North America, motor-vehicle collisions are a leading cause of death. In fact, they're the No. 1 cause of death among

KILLER PHONE & KILLER ALCOHOL

According to a *New England Journal of Medicine* study, a cellular phone can make you as dangerous behind the wheel as drinking lots of alcohol! In the past, a drunk driver was seen as a joke, and it took many years before a drunk driver was seen as a potential killer. For how many years will we allow drivers to use cellular phones?

According to American statistics, if 10 per cent of American cars carry a cellular phone the extra cost to society will be about $US3 billion per year.

children and young adults. Each year, about 2 per cent of the population in North America are involved in a car accident — 25 per cent of those people will have a temporary disability, 10 per cent will end up in hospital, while 1 per cent will die. Anything that leads to an increase in car accidents has to be a bad thing.

One study looked at drivers making a phone call, and at how far they turned the steering wheel. The researchers found that while people were talking on the phone, the size of their rotation of the steering wheel more than doubled. In other words, their steering was less accurate and they wobbled across the road.

Another study, reported in the *New England Journal of Medicine*, looked at 700 drivers who had cellular phones and who also were involved in a motor-vehicle accident. The researchers found that if you were driving the car while using a cellular phone, you increased the risk of a car accident by about *four times*.

But the surprising thing was that they couldn't really see any difference between hands-free units and hand-held units. The sample size of people wasn't quite large enough to work out whether there really was a statistical difference, and they really do need a follow-up study with bigger numbers. But it could well be that the mere act of having a conversation with someone you can't see will reduce your concentration enough to make you more likely to have a car accident.

So steer clear of cellular phones when driving, and keep your hands on the wheel and your mind on the job.

Colon injuries

My bathtime reading has included a few interesting medical papers related to the bowels and to death. The first one was called 'Fatal colonic explosion during colonoscopic polypectomy'.

'Fatal' means that the patient died. 'Colonic' means that the explosion happened somewhere in the patient's colon, or bowel. 'Colonoscopic' means that the doctors were using a viewing scope to look for any abnormalities in the patient's colon.

The colonoscope is a flexible tube, a few centimetres in diameter and about a metre long. It has a light, viewing lenses with variable magnification, and little wire loops that cut through flesh and seal it as they go.

'Polypectomy' means to remove polyps. Polyps look like a punching ball, and are abnormal growths that sprout out from the normal bowel wall on a thin stem — anything from a millimetre to a few centimetres in size. Medical doctors try to remove them from the colon because some polyps, but not all, turn into cancers.

So 'Fatal colonic explosion during colonoscopic polypectomy' means that while the doctors were examining a patient's colon, and removing any potentially cancerous polyps they found, there was an explosion that killed the patient.

How did this happen?

Well, the doctors needed a good look at the inside wall of the colon. So they cleaned the inside of the bowel, and they inflated the colon with air so the colon was not all crinkled up and possibly hiding a polyp in a fold.

Back in 1979, they used a chemical called mannitol to clean the bowel. Mannitol sucks water from the blood; this water fills the bowel and the contents are then washed out through the back passage. Some people have bacteria in their colon that will eat some of the mannitol and turn it into hydrogen. Hydrogen, in concentrations of between 4 and 74 per cent, will explode when mixed with room air.

In the case of this patient, the doctors had found a polyp and had looped their wire around it. To quote from their paper:

'A coagulating current was used . . . After 8 to 10 seconds of current passage, there was an explosion which was audible in the endoscopy room, the patient jerked upwards off the endoscopy table, and the colonoscope was completely ejected.'

Obviously, a spark from the cutting wire had exploded the hydrogen gas in this unlucky patient. The patient immediately went into shock, and despite a massive blood transfusion of some 45 bags of blood, died shortly afterwards. An autopsy showed dozens of sites of bleeding inside the patient's abdomen.

Today, as a result of that event in 1979, mannitol is no longer used to wash out the bowel. It is now the official policy of the French Endoscopy Society always to blow an inert gas (such as carbon dioxide) into the bowel during these procedures, to make sure such explosions cannot happen.

Coffee enemas

Another interesting paper was entitled 'Death related to coffee enemas'. It's well known that the British Royal Family enjoys sipping tea, but it's not so well known that certain members of the British Royal Family believe that there are enormous health benefits associated with having a litre-or-so of coffee flushed in, and out of, their back passage. But coffee enemas have killed!

A 46-year-old woman was suffering from gall bladder disease. She refused conventional medical treatment, and instead, asked her naturopath to give her coffee enemas, as frequently as three or four per hour! Her pain decreased, and she claimed that she had passed several gallstones in her bowel motions — however, she continued having the coffee enemas at the rate of one every hour.

NO EXPLOSIONS TODAY

Today, when patients are prepared for a colonoscopy, they are given a mix of a low-residue diet, oral laxatives and an enema. These procedures reduce the concentrations of the explosive methane and hydrogen gases down to about one-twentieth of the minimum level needed for explosion.

Suddenly she had a convulsive fit, followed by two more. The next morning she had a fourth fit, which was immediately followed by her heart and lungs stopping. Doctors managed to crank up her heart and lungs again, but she remained in a coma until she died 12 days later.

The second case involved a 37-year-old woman with breast cancer, who had had her right breast removed. During surgery, the surgeons found that the cancer had spread to many of her lymph nodes throughout her body. According to the report, 'she refused chemotherapy after the surgery. About six months later, she noted nodules swelling up in the right side of her neck [and around her breast].' She was started on immunotherapy, but didn't respond well. Two months later, she started on chemotherapy. Another two months later, her doctors recommended radiation therapy, which she refused. She also discontinued the chemotherapy.

One month later, she attended a clinic where she received a special diet. Part of her treatment involved coffee enemas four times a day, using 0.95 of a litre of coffee per enema. She continued the treatment for about six weeks, and then began to feel dizzy and suffer from pains in her chest and abdomen. She died shortly afterwards. The autopsy revealed that the cancers in her body were probably not the cause of her death.

In each case, the autopsy revealed that the most likely cause of death was a massive loss of potassium. The body is very sensitive to the levels of potassium in the blood-stream. The coffee enema had a strong osmotic (or sucking) effect, and pulled potassium out of the women's blood-streams, into their bowels, and away.

I guess the bottom line is that too much coffee is bad for you, no matter which end you put it in!

Fainting fans

Another modern disease is described in a paper from the *New England Journal of Medicine*, entitled 'Mass fainting at rock concerts'. In this particular case, the doctors interviewed 40 of the 400 people who had passed out at a concert in Germany put on by New Kids On The Block. They were all girls, and they were all between 11 and 17 years of age. It turned out that there were two quite distinct medical conditions here.

Some 40 per cent of the young women actually went into unconsciousness. They all reported that on the previous night they had been so excited that they couldn't sleep, and on the day of the concert they had been so excited they couldn't eat. They also had spent a long time standing up in the arena. Just before they fainted, they had either been squashed by the crowd or had been screaming really loudly. Being squashed, and screaming, increases the pressure inside your chest cavity. When the pressure inside the chest goes up, the amount of blood returning to the heart drops and so does the amount of blood leaving the heart. When the blood pressure drops below a certain level for a few moments, not enough blood gets to the brain and you faint.

But 60 per cent of the young women had a different syndrome. The medical name for what they suffered is a hyperventilation or panic attack. Even though they'd fallen to the ground, they remained completely alert. Part of hyperventilating is that you actually close down some of the blood vessels in the brain and, once again, with not enough blood, part of the brain temporarily goes into the sleep mode.

In their paper, Drs Thomas Lempert and Martain Bauer have pretty well explained 'The multi-factorial pathophysiology of rock concert syncope' — which is medical talk for the many causes that make people faint at rock concerts. They've also come up with some very good suggestions: anybody going to a rock concert should 'sleep, eat, sit, keep cool, and stay out of the crowd'. But what's the point of being a teenager and going to a rock concert if you're going to do all that boring stuff?

Healing hole in the head

Surgery has been around for thousands of years. In fact, surgery has been called 'the peaceful use of wounds'. As far as we know, the earliest surgery was done about 12,000 years ago, and it was trephination — the cutting of a hole into the skull. This surgery was practised by many different societies around the world — and the amazing thing is that in New Guinea in the 1800s, the success rate of this operation was much higher than in the operating theatres of Europe!

Trephination has been done in ancient Peru, in prehistoric Sardinia, in the mountains of Algeria, and in many parts of the Pacific Rim, including New Guinea and Fiji.

In the 1870s, the Tolai medicine men around Rabaul in New Britain had a success rate around 70 per cent, while patients of Guy's Hospital in London had only a 25 per cent chance of surviving the operation!

The main reason for trephination, or cutting into the skull, was to repair an injury received in battle. For example, in Fiji, the sling-stone — a hard, smoothly rounded stone the size of a hen's egg — was a fearsome weapon. An expert could hit a man's head at 300 metres. Occasionally, the force of the impact was great enough to explode the man's head. But usually, the sling-stone would crack the skull and cause bleeding inside.

Surgery would have to be done immediately if there was to be any chance of saving the patient. First, the surgeon would wash his hands in the milk of an unripe coconut. This was a very good start to the operation. Not only is this milk sterile (that is, it has no germs in it), it is very close to the chemical mix of our bodily fluids.

In most cases, the patient was still unconscious from the injury, so there were no complaints when the surgeon would make a triangular cut around the site of the head injury. In New Britain, the medicine men used sharpened bamboo, the teeth of sharks or even sharp shells as a knife. An assistant would pull on the patient's hair to keep it out of the operating area. Two women would be given the job of supplying coconut milk — one to continually open coconuts, the other to slowly and continuously pour the milk over the patient's wound and the surgeon's hands.

Once the surgeon had exposed the hole in the bone that was caused by the injury, he would work away at it until he made it oval. Air was blown into the wound via a hollow pipe to expose any fragments of bone. Such fragments would be picked out with tweezers made from bamboo. Any damaged brain tissue would be scooped out.

The hole would then be plugged up with some bark and the skin closed over it. The skin was sewn together with threads from banana fibres, using needles made from the wing bones of the flying fox. Finally, the wound was covered with banana leaves and a mixture of chewed-up betel nut and banana flower.

Three days after the operation, this dressing was taken off. If the wound was clean — well and good. But occasionally it would be infected, and the pus then would be let out and the wound cleaned.

But how come European hospitals had such low survival rates from this operation? First, even back then, over-crowding in Europe had led to the evolution of nasty bacteria, which had not yet arisen in the so-called 'primitive' parts of the world. Second, the sterile coconut milk was cleaner than anything then used in Europe, and so the wound was cleaner.

Coconut milk has roughly the same sodium and potassium levels as bodily fluids. Even today, it can be used to resuscitate dehydrated children suffering from diarrhoea. In fact, during the Vietnam War, the lives of many injured soldiers were saved when they were given this coconut milk directly into their veins after the medicos had run out of blood and other transfusible fluids.

Maybe Carmen Miranda had the right idea when she wore all that fruit on her head!

REFERENCES

Bigard, Marc-Andre *et al.*, 'Fatal colonic explosion during colonoscopic polypectomy', *Gastroenterology*, vol. 77, no. 6, 1979, pp. 1307–1310.

Cooter, Robert, 'The Royal Flying Doctor Service', *Australian Family Physician*, vol. 26, no. 1, January 1997, pp. 58–61.

Eisele, John W. & Reay, Donald T., 'Death related to coffee enemas', *Journal of the American Medical Association*, vol. 244, no. 14, 3 October 1980, pp. 1608–1609.

Lempert, Thomas & Bauer, Martin, 'Mass fainting at rock concerts', *New England Journal of Medicine*, vol. 332, no. 25, 1995, p. 1721.

Majno, Guido, *The Healing Hand: Man and Wound in the Ancient World*, Harvard University Press, 1975, pp. 24–28, 166–169, 197.

Mourad, Jean-Jacques, 'Carotid-artery dissection after a prolonged telephone call', *New England Journal of Medicine*, vol. 336, no. 7, 13 February 1997.

New Scientist, no. 1989, 5 August 1995, p. 64.

Redelmeier, Donald A. & Tibdhirani, Robert J., 'Association between cellular-telephone calls and motor vehicle collisions', *New England Journal of Medicine*, vol. 336, 13 February 1997, pp. 453–458.

Stewart, Betty, 'Brain surgery in the South Pacific', *Medical Observer*, 17 March 1995, pp. 74–75.

Westmore, Ann, 'Fall leads to artery dissection', *Australian Doctor Weekly*, 16 May 1997, p. 9.

Bioinformatics & Immortality

This is probably the first science story that I have written that has a lot of future prediction and speculation in it. Usually I keep to the amazing-but-true facts, and allow myself only a very small speculation at the end of the story. But as this story deals with such a big topic — effective immortality — I feel compelled to rush into print. I could be wrong, but I don't think so.

Bioinformatics, a scientific revolution

Very quietly, we're heading for a new scientific revolution. This revolution will let us grow arms for people who lost an arm in an accident, bring our skin back to a youthful teenage vigour, and get rid of osteoporosis. Thanks to this new technology, some of us might be in the first generation to live forever or the last generation to die!

Even though this revolution won't fully erupt until around 2003, the industry is already paying people in this field six-figure salaries.

There are many names for this new field of study. They include genomics, the Human Genome Project, genetic engineering and

WHAT IS DNA?

D stands for deoxyribose. Deoxyribose is a type of sugar. There are three billion deoxyriboses making up one of the side rails of the ladder of the DNA molecule, and another 3 billion deoxyriboses making up the other side rail.

The NA part of DNA stands for nucleic acid. The nucleic acids include the four bases: adenine, cytosine, thymine and guanine. These names are usually shortened to A, C, T and G. (There's another type of base called uracil, but that's not important in this explanation.)

A and T will always join together (so you get AT and TA rungs), and C and G will always join together (to make CG and GC rungs). Normally, you identify the rungs by the first letter. So you can have only four types of rungs in the ladder of life: A, C, G or T.

Three physicists, James D. Watson, Francis Crick and Maurice Wilkins, shared the Nobel Prize in 1962 for the great discovery that a set of three rungs contain all the information to tell the other biological machinery in the cell to make an amino acid. (Wilkins's colleague Rosalind Franklin did not share the Nobel Prize with them — some scientists think this was unfair.)

So the set of rungs AAA will make one amino acid, while the rungs AAC will make a different amino acid. (There are 64 possible different rungs, but only 20 different amino acids that the DNA makes. So sometimes, two or more different amino acids can make the same amino acid.) When amino acids join together, they make proteins.

bioinformatics. But what's really important about this field of study, regardless of what we call it, is that *for the first time in human history we can actually change our human DNA.*

The human cell has a cell membrane around it, and it's a few microns (millionths of a metre) in diameter. Inside the cell are all sorts of biological machines: mitochondria, which provide us with energy; centrioles, which help in the division of the cell when it splits into two; and various other pieces of biological machinery that make proteins. The proteins are the key. These proteins can either be used internally by the cell, or exported outside the cell for use somewhere else in the body (such as thyroid hormone, insulin, and growth hormone). And running the show, right in the middle, is the DNA all wrapped up inside the nucleus. The DNA controls the manufacture of proteins.

The DNA looks like a ladder — with two side rails and a whole bunch of 'rungs' (they're also called 'codons', 'base pairs' or 'letters'). In fact, there are about three billion rungs (but only four different types of rungs). If it were all stretched out, your human DNA would be about 2 to 5 metres long. But the DNA has been all coiled up to fit inside the nucleus of a cell only a few microns across.

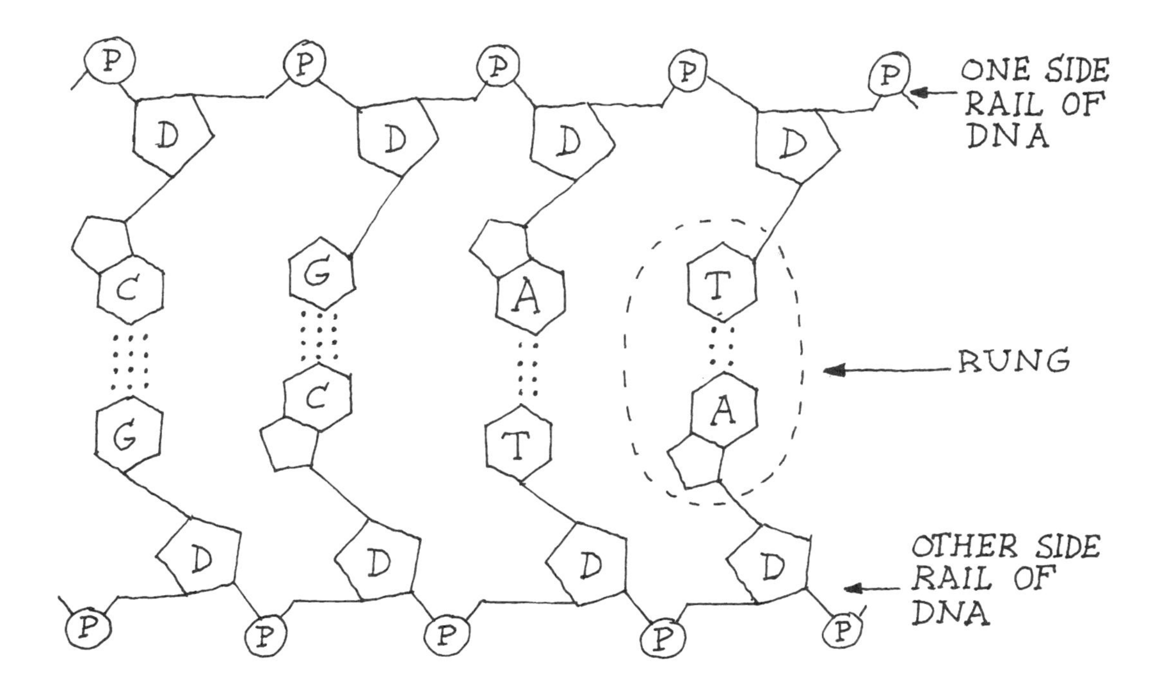

D= DEOXYRIBOSE

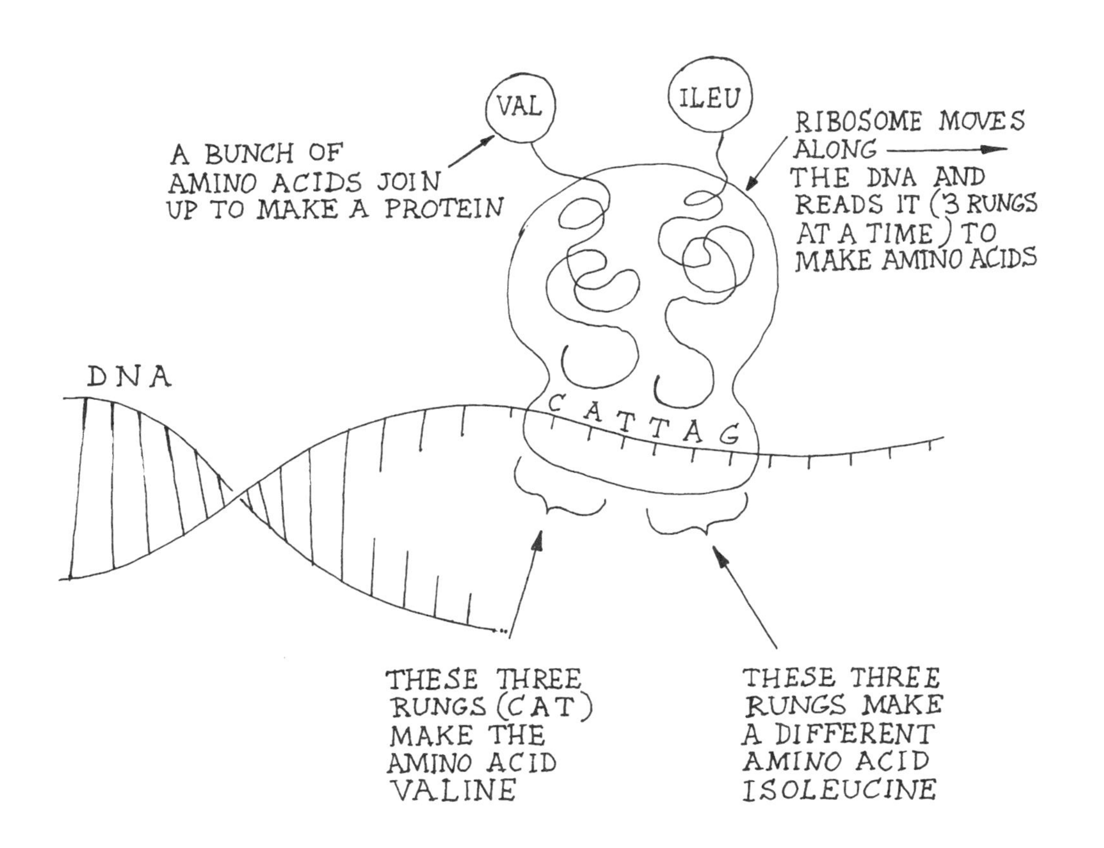

The DNA controls the making of all the different cells in the body. If one small part of the DNA is activated, it triggers the growth of cells that are beautifully suited to dumping carbon dioxide and picking up oxygen — the cells of the lungs.

If you activate another part of the DNA, you make a structure that is self-repairing and, weight for weight, stronger than steel — bone.

The orthopaedic surgeons call bone 'the prince of tissues', because it heals so beautifully. But bone has another magnificent characteristic — it is continually rebuilding itself. Inside the bone are two classes of cells (osteoclasts and osteoblasts) that are continually destroying and rebuilding your bones. Over a period of some seven or eight years, all of the calcium atoms in your bones will leave your body via your bowel and be replaced by new calcium atoms that come in through your mouth via the food you eat.

It's very important to realise that your bones rebuild themselves 'on the run', while you still walk on them. At no stage do you have to go to a hospital to have all of your 206-or-so bones removed and replaced by new ones.

Re-growing on the run

When we will have learnt to read, write and activate the DNA as we wish, we will be able to re-grow any part of the body we want. We will be able to grow, on the run, a kidney or an eye or a bone.

Some people talk about the evils of genetic engineering, where a clone of you will be grown so that it can be later 'harvested' for a replacement kidney. (For some body parts, such as the heart, of course the 'harvesting' would kill the clone.) This is a ridiculous way to get a new kidney. In the pathway that they envisage, your DNA will replace the native DNA of an already fertilised egg, which will then be implanted in a uterus. The egg then spends nine months growing into a baby and another 20 years growing into an adult, at which time the kidney is supposedly removed and the adult somehow 'got rid of'. This is murder. It is also extremely stupid — why wait some 20 years for your transplant kidney? You could be dead well before your replacement kidney has grown in your clone.

Surely it would be much easier to learn the trick that bone uses, and re-grow your diseased kidney into a healthy kidney *on the run*. This would happen over the period of a few months.

The liver knows a different version of the same trick as well. In the Greek legend, Prometheus was the god who gave the gift of fire to the humans. Zeus, the chief god, was angered by this, so he punished Prometheus by chaining him to a rock and having an eagle come every day to feed upon his liver. Each night, his liver would re-grow. We don't know how the Greeks knew this, but if you remove one-third of the liver (say, for a transplant) the remaining two-thirds will re-grow.

The blueprint of life

When we manipulate the DNA, we control the blueprint of life.

At its simplest level, DNA makes proteins. These proteins can sometimes be enzymes. Enzymes speed up chemical reactions. Depending on what you want an enzyme to do, it can either eat the food

BLACK PYGMY & WHITE CAUCASIAN

There is very little difference between a black Pygmy !Kung woman and a tall, blond male Scandinavian, as far as their DNA is concerned.

First, the male–female difference. As far as the DNA is concerned, the only difference between males and females is that females have an X chromosome (with four arms radiating out from a central point), while males have a Y chromosome (with three arms radiating out from a central point). So females are the luxury model with all of the options, while males are the cheap economy model missing a few options. Even so, the difference is very small.

The difference in colour is also a very minor point. Your skin colour is due to the amount of melanin present in your skin. This melanin comes from cells called melanocytes. All humans have roughly the same number of melanocytes per square centimetre of skin. It's the *activity* of the melanocytes that makes the difference between a fair-skinned and a dark-skinned person. In the fair-skinned Scandinavian the melanocytes are mostly inactive, while in the dark-skinned African they are mostly very active.

The third main difference is height. But once again, this is only a very minor consideration. Your final height depends on how much growth hormone is pumped into your body at different times in your life. The growth hormone comes from the pituitary gland in your brain. Both the Pygmy and the tall Scandinavian have functioning pituitary glands. The difference is that the Pygmy's pituitary gland, following instructions from the DNA, pumps out much smaller amounts of growth hormones at different times.

One lesson that we have learnt from genetic engineering so far is that the differences between us are much less than our similarities.

stains out of your clothing or it can speed up the chemical reactions to make flesh and skin, blood and bone, brain and liver, and everything else you need for a new baby.

Today, you are a human made up from about 10 trillion cells, but when you were first conceived you were just one single cell. Along the way, you grew two arms, two legs, two eyes, one liver, and so on. The blueprints, or instructions, for all your body parts were contained in the DNA in that single fertilised cell. These instructions still exist in every cell in your body, except for your red blood cells. (This might be because red blood cells have only the simple job of delivering oxygen, and so they don't need much brain power.)

Once we learn to work the DNA, we can grow an arm!

Consider the all-too-familiar scene that you see on the roads — the driver with one arm illegally hanging out of the car window. Every now and then, the arm gets scraped off by a vehicle going the other way. If the arm gets badly damaged, microsurgery won't work, and the driver

RE-GROW LUKE'S ARM!

Much as I love the *Star Wars* trilogy, I believe they were wrong when they gave Luke Skywalker an electromechanical arm to replace the one that was chopped off. This technology to grow a biological replacement right where the old one used to be should be available within 20 years at the most — a bit longer than it took to get 12 humans onto the surface of the Moon.

will have to wear an artificial arm for the rest of his or her life. But a long time ago, in the uterus, the driver once grew an arm — and the instructions on how to grow an arm are still there in the driver's DNA. Wouldn't it be wonderful if we could find those instructions in the DNA, and activate them, so that the driver could grow another arm! It could probably reach full size in less than a year. After all, if the salamander can grow a new limb, why can't we?

It's not impossible to grow new body parts — our scientists have already done it. A fruit fly made the front cover of *Science* in October 1995. Scientists studying the common fruit fly found the specific section of its DNA that controlled the growth of its eyes. They learnt how to switch on the growth of eyes. This fruit fly made the front cover of *Science* because it had 18 eyes!

It had eyes all over its body. The eyes that were grown by human intervention were not as perfect as the eyes that were grown by the untouched DNA of the fruit fly. We can't control how good these 'extra' eyes are, and we can't control where the eyes will go — yet. One of these eyes even ended up on a knee of the fruit fly!

It's early days yet, and the genetic scientists have a lot to learn before we can grow a human arm as easily and reliably as a salamander can re-grow an amputated limb. We would want the arm to grow where the lost arm was, not from the back of the neck, or the knee, or anywhere else!

Mapping

The entire human DNA has some three billion letters all jumbled up. There are only four different types of letters (A, C, G and T), but there are an awful lot of them. How will we make sense of all this information?

Marshall McLuhan, who wrote *The Medium is the Message*, gave us a hint when he said: 'The map defines the territory.' There are only five words in that short sentence, but in that particular combination they give a very powerful message.

Consider Dmitri Mendeleyev (or Mendeleev) and his organisation of the elements and the Periodic Law.

In the 1870s, he, like the rest of the chemists in the world, had access to a list of the then-known elements. He grouped them in a table with 18 columns. He suddenly could see that there were three gaps in that list of elements. Then, in 1871, he made two predictions. His first prediction was that three elements were missing from the table of elements — not two elements, not four elements, but three elements. He was right! His second prediction was that each element that was missing would have similar properties to those elements that were around it. So, on the left-hand side of the Periodic Table of the Elements, the missing element between calcium and titanium would have similar properties to calcium and titanium (such as boiling point, melting point and appearance). Once again, for each of the three missing elements, he was right.

He could not have made this prediction without the map! In other words, the map gave him extra knowledge.

To 'map' a part of the DNA is to know that a particular segment goes, for example, AACGTAGC. The goal of the Human Genome Project is to map the whole of the human DNA. The project began in a vague kind of way back in the 1970s, when scientists realised that it would be soon possible to know exactly what every rung in the DNA was. The project gradually picked up momentum as various countries became involved. The Human Genome Organisation (HUGO) was started in 1989 to co-ordinate the mapping on an international level. The current president of HUGO is Professor Grant Sutherland of the Women's and Children's Hospital in Adelaide, South Australia.

At the current rate of progress, we should have mapped all of the DNA by somewhere around 2003. By then, we should know exactly which small section of DNA holds the instructions to grow a new arm. That section might already be sitting, unrecognised, in a site somewhere on the Web.

Even though the whole of the human DNA won't be mapped until around 2003, the job opportunities in bioinformatics and genetic engineering are enormous. The back pages of *Nature* and *Science* carry advertisements every week. According to the September 1996 issue of *Wired*, graduates in bioinformatics are getting up to US$100,000 per year.

The Human Genome Project deals with one of the Big Questions of all time, so there'll be lots of employment for lots of people for a long time. There'll be jobs for physicists, molecular biologists, biologists, biochemists and people from dozens of other fields — but one of the huge job-opportunity areas will be in the field of computing.

If you were to put all three billion bits of information (in other words, the three billion letters which are one of A, C, G or T) into a book and read them for eight hours a day, it would take a few decades of your working life to get through the book! When a person missing an arm comes into casualty, he or she won't want a six-hour discussion about genetic engineering — what the patient will want will be a bottle of retrovirus called Grow-New-Arm-O, which he or she will drink, and which will start the growth of an arm. This project will simply be impossible without assistance from the computer scientists.

Skin and bone repairs

My prediction is that one of the very first products to come to market will be a retrovirus to repair your wrinkled skin. Your skin wrinkles because, with time, the production of elastin and collagen in your skin slows down. You might think that trying to have good skin would be a trivial pursuit, but then, you've probably never tried to get blood from an 80-year-old person.

Babies have magnificent skin, so delicious that you want to kiss them a lot, teenagers have pretty good skin, and even people in their 50s and 60s have pretty reasonable skin. But people in their 80s and 90s often have paper-thin skin, which bruises easily on any sort of contact. Wouldn't it be nice to be able to go to your local chemist and buy a bottle of retrovirus called Grow-New-Skin-O, which would then switch on the production of collagen and elastin in your skin, and then for the rest of your life you would have magnificent teenage-quality skin? (Of course, the retrovirus would have to switch on production of collagen and elastin *only* in your skin — you wouldn't want to grow collagen in your eyeball, brain, muscle, pancreas or anywhere else in your body.)

Another good body system to improve would be bone. In some of the wealthier parts of the world, the main reason that an older woman is admitted to a teaching hospital is because she has osteoporosis, and when she tripped and fell, she broke the neck of her thigh bone. If she's lucky, her osteoporosis is mild enough so that the neck of her thigh will repair and she can enter the community again. If she's unlucky, the neck of her thigh will never repair and she could be stuck with forever living in a wheelchair or a nursing home.

Five stages to genetic engineering

As I see it, there are five stages to this science we call genetic engineering.

The first stage of genetic engineering is to *modify a living creature for OUR benefit.* In the bad old days, diabetics had to inject themselves with insulin made from the pancreas of pigs. Today they can buy, from their local chemist, genuine human insulin. This insulin is not made by squashing up human beings, but by genetically modified bacteria that live in vats and make the insulin hormone for our use. We have already successfully done this. (Of course, further down the track, the diabetic would simply grow a new pancreas.)

The second stage of genetic engineering is to *modify a creature for ITS benefit.* Tobacco plants are often attacked by insects. By a coincidence, there was living many thousands of kilometres away from the tobacco plants, a fungus that made a chemical which would kill these insects. The genetic engineers married the fungus and the tobacco plant together, so now there is a tobacco plant that makes its own insecticide. This is good for the tobacco plant, because it no longer gets attacked by the insects. But because it has this chemical in it, it tastes bad and doesn't smoke very well either. Even worse, it encourages resistance to the insecticide amongst the insects, because this chemical is continually being made by the tobacco plants. As we get more skilful, we could make this tobacco insecticide work a little bit like our own human immune system, which switches on only when it is attacked.

The third stage of genetic engineering is to *modify human beings, to get rid of a few nasty diseases.* There is a very rare child-hood disease called Severe Combined Immunodeficiency Disease (SCID). These kids have virtually no immune system, and rarely survive past their first birthday without treatment. There are about five different types of SCID, but about half of the kids suffer from just one type, which is caused by one single missing enzyme (adenosine deaminase). Our scientists have been able to genetically engineer a retrovirus, with which they then infect these children. The retrovirus then switches

GENETIC WEAPONS

Genetic engineering can be used as a force for good, and also as a force for evil.

Imagine you are engineering a virus. If you choose a virus that is as hard to catch as HIV, not many people would catch it. But if you choose something like the influenza virus, it could be easily transmitted by sneezing and it could sweep around the world within a few years.

If you genetically engineered the Ebola virus so that it had a two-week delay before it became active, and was as infective as the influenza virus, you would have a truly nasty weapon. You could further genetically engineer it so that it would become active only in somebody who had the particular characteristics that you personally hated — such as skin colour, eye colour, hair colour, height, or even sex.

on the production of this missing enzyme in some (but not all) of the kids for a little while (but not for the rest of their lives) to a limited extent (but not so that they have a completely functioning immune system). What we would like to do is give them an immune system as good as everybody else's by giving them a single dose of the retrovirus, which would last them for the rest of their lives — but in the late 1990s, we haven't even got the complete map of the human DNA yet.

The fourth stage of genetic engineering involves *improving parts or systems of the body so that we could live 'forever'*. Practically all the systems of body slowly lose function with time — the skin, the bone, the lens in the eye (which stiffens so that you gradually lose the ability to focus), the muscles, the kidneys, the heart and blood vessels, the liver, and so on. Once all of these have been rejuvenated, by 'fooling around' with the DNA, we're looking at a healthy body that can behave like a teenage body for 500 to 1,000 years.

This is a short-term effective immortality — so people living today could be the first generation to live 'forever', or be the last generation to die.

The fifth stage of genetic engineering would be to *modify human DNA so that (for example) we could live on a planet like Mars with only a simple garment — not a spacesuit — to protect us.* We would have to survive sub-zero temperatures (without freezing) and the incredibly thin atmosphere of carbon dioxide (with a pressure about 160 times less than our own).

This level of skill with genetic engineering should be achieved within three-quarters of a century — the same time it took to go from the first human flight in an aeroplane to the first human landing on the Moon. How long before the first natural regrowth of an arm or leg?

Technology = possibility

Growing an arm might seem impossible, but once upon a time, flying also seemed impossible.

My father was born in 1900, and died in 1981. Around 1905, the Wright Brothers

EVERY DAY, TWO OUT OF THREE GO HUNGRY

In the middle of our excitement about bioinformatics and our self-congratulation at being able to move into a brave new world, we should nevertheless realise that two-thirds of the people on our planet are desperately poor.

In much of the world, the total annual income earned to run a family is less than the pocket money of teenagers in wealthy countries. The poor countries don't have enough food, health care, doctors, education, clean drinking water and so on. In most cases, these countries are poor because the wealthy countries are ripping them off. In many cases, the amount of money paid in 'aid' to these poor countries is one-tenth of what these poor countries pay in bank interest to the wealthy countries!

Fred Hollows, that great Australian doctor and humanitarian, once said to me: 'The presence of a poor person anywhere else in the world lessens me as a person.'

had the first reliable and safe controlled flight. By 1969, we had gone to the Moon and back. Imagine how impossible it would have seemed back in 1900, if somebody had predicted that in the lifetime of a baby born in 1900, not only would the human race learn controlled flying, but also that we would send people to the Moon and back.

The reason that it was possible for this to happen in such a short time as one person's life span was because all the necessary technology was in place — the technology of gliding, and a light and powerful engine.

In the same way, by around 2003 all the technology will be in place to deal with the human DNA.

REFERENCES

Blaese, R.M. & Culver, K.W., 'Gene therapy for primary immunodeficiency disease', *Immunodeficiency Review*, vol. 3, no. 329, 1992.

Cohen, Jon, 'The genomics gamble', *Science*, vol. 275, 7 February 1997, pp. 767–772.

'Frontiers in medicine', *Science*, vol. 276, 4 April 1997, pp. 59–87.

Horton, Brendan, 'Going to work in genes catches on', *Nature*, vol. 383, 24 October 1996, pp. 739–743.

Littlejohn, Tim G., 'Bioinformatics: the essential ingredient', *Today's Life Science*, June 1996, pp. 28–33.

Westmore, Ann, 'Disease clues in the genes', *Australian Doctor Weekly*, 16 May 1997, pp. 39–42.

Williams, D.A., Lemischka, I.R., Nathan, D.G., *et al.*, 'Introduction of new genetic material into pluripotent hematopoietic stem cells of the mouse', *Nature*, vol. 310, 1984, p. 476.